Mental Exercise for Dogs

Transforming Your Canine's Behavior and Building a Stronger Bond through Innovative Mental Activities Using Simple and Effective Games and Exercises

Laurel Marsh

Mental Exercise for Dogs
© Copyright 2023 by Laurel Marsh
All rights reserved

TABLE OF CONTENTS

Laurel Marsh

CHAPTER 1

UNDERSTANDING THE CANINE MIND

The Importance of Mental Stimulation for Dogs

Dogs thrive when their minds are actively engaged. Mental enrichment is just as critical to a dog's wellbeing as physical exercise and a nutritious diet. An understimulated dog is more likely to develop problem behaviors like destructiveness, hyperactivity, or aggression. Mental stimulation also strengthens the bond between owner and pet, as training and puzzle toys provide opportunities for positive interaction.

The canine brain requires regular challenges and new experiences to stay healthy. Dogs are intelligent, social animals evolved to solve problems and cooperate with humans during activities like herding, hunting, and guarding. Modern life often does not provide enough opportunities to tap into these natural instincts. Mental exercise activates neural pathways in the brain, building cognitive skills and helping dogs focus. Puppies especially need mental stimulation during crucial development stages. Exposing them to new sights, sounds, and challenges primes them for training and temperament tests.

Mentally stimulating a dog does not require complicated equipment or agility courses. Simple games of hiding treats and having a dog "hunt" for them activates their natural foraging skills. Introducing new toys, walks in different locations, learning tricks, and social visits all provide cognitive enrichment. Puzzles and interactive feeders also encourage dogs to use their problem-solving abilities. Rotating different activities prevents boredom from repetition. Providing various "jobs" tailored to a dog's breed gives them an outlet for their energy and abilities.

The benefits of brain games extend beyond tiring a dog out. Activities strengthen the understanding between handler and dog, establishing communication and trust. Dogs learn by associating cues with actions, reading human body language, and positive reinforcement through rewards and praise. This builds confidence in their abilities. Owners get insight into how their dog thinks and learns. Puzzle toys are especially useful for high-energy dogs as an alternative to destructive chewing. They satisfy natural behaviors in a healthy way.

Mental exercise also slows age-related cognitive decline. As dogs mature, their brains undergo changes similar to dementia and Alzheimer's in humans. Keeping them mentally active with training, toys, and social interaction preserves neural connections. This may delay or reduce behavioral changes associated with canine cognitive dysfunction syndrome. Senior dogs who stay active and responsive tend to have better quality of life.

Signs a dog needs more mental stimulation include hyperactivity, chewing, barking, attention-seeking, and other anxiety-driven behaviors. Providing an outlet for energy reduces stress. Dogs also benefit from establishing daily routines with built-in activities. Schedules create stability while preventing monotonous repetition. Incorporating both training and free play prevents frustration. Alternating high-focus activities with relaxing downtime ensures a balanced approach.

Introducing brain games and training gradually is important, especially for shy or fearful dogs. Activities should build confidence, not add more stress. Keep initial sessions short and positive, using rewards to motivate without pressure. Be patient and let the dog set the pace. Seek guidance from a trainer for dogs with specific behavioral issues or sensitivities. The right approach can make mental exercise an enriching experience.

In summary, mental stimulation is a vital component of good canine health, engaging natural instincts and abilities. Brain games reinforce the pet-owner bond through communication, trust, and problem solving skills. Mental exercise benefits dogs of all ages by building cognitive resilience and preventing boredom-related behaviors. Adjusting activities to match each dog's abilities and needs ensures they stay challenged.

Just like physical activity, regular mental enrichment engages a dog's body and mind for better wellbeing.

How Dogs Think and Learn

Dogs have complex cognitive abilities that allow them to learn and adapt to their environments. Their thinking processes and learning styles have been shaped by evolution to ensure survival. While dogs are intelligent animals, their cognition differs in key ways from human cognition.

When it comes to perception, dogs rely heavily on their senses of smell, hearing, and sight. Their sense of smell is exceptionally strong, with up to 300 million scent receptors compared to a human's 5-6 million. This allows dogs to gain crucial information from smells in their surroundings. Hearing is also keen in dogs, enabling them to detect a wide range of noises. They can hear frequencies up to 45 kHz, while humans hear up to 23 kHz. Dogs have less visual acuity than humans, but their vision is well-adapted for detecting movement. Overall, dogs build an understanding of their world based on sensory input.

In terms of intelligence, dogs have excellent associative learning abilities. This means they form associations between stimuli, responses, and outcomes. For example, they associate the sound of a bag crinkling with getting a treat. Dogs also have good capacity for spatial learning to navigate environments. However, their ability to infer causal relationships is more limited compared to humans. While dogs may learn that performing a trick earns a reward, they don't assign as much meaning to that causal link.

When it comes to concepts, dogs form categories to understand their world. They can group objects based on common characteristics. For instance, they recognize that all balls, regardless of differences, belong to the same category of objects that can be fetched. Dogs are also good at learning routines based on time or order of events. If fed at fixed times daily, they anticipate mealtimes. Their concept of time itself is limited though.

In terms of social cognition, dogs excel at reading human cues like gestures and facial expressions. Through the process of domestication, they have become adept at understanding human communication signals. Their ability to perspective take and infer others' mental states, however, remains rudimentary compared to humans.

Problem solving skills in dogs depend on the breed. Working dogs and herding breeds tend to have better independent problem-solving abilities. Scent hounds follow smell trails, while sight hounds use vision to pursue prey. Retrievers persistently solve puzzles to earn rewards. Terriers excel at making decisions and assessing risk. Overall, a dog's genetics and selective breeding influence their cognitive style.

Now let's explore some key theories on how dogs learn. One major theory is classical conditioning, based on the work of Ivan Pavlov. This involves associating a neutral stimulus with a reflex response. The classic example is Pavlov's dogs salivating at the sound of a bell associated with getting fed. The bell's sound provoked the involuntary reflex of salivation. Operant conditioning is another model identified by B.F. Skinner. This posits that behaviors are shaped by their consequences. Dogs repeat actions that are positively reinforced with rewards.

Cognitive learning theory is also relevant to dogs. This states that mental processes like perception, thought, memory and motivation influence learning. Insights into dog cognition can help optimize training techniques. Information processing theory focuses on how dogs encode, process and retrieve information. Understanding their mental processes facilitates learning. Social learning theory highlights how dogs can learn by observing other dogs or people, without direct reinforcement.

Overall, several factors impact how dogs learn effectively. Breed differences lead to variation in cognitive styles that may require tailored training approaches. A dog's life experiences also shape their learning abilities. Early socialization and exposure to variety helps dogs learn more flexibly as adults. Motivation levels also affect learning, with rewards like food or toys enhancing motivation. Following positive reinforcement principles is key, as is clear communication. Patience, consistency and repetition are likewise integral when training dogs.

In summary, appreciating dog cognition and theories on learning allows humans to unlock dogs' potential. We can leverage their natural abilities and conditioned behaviors to teach them successfully. Studies continue elucidating how dogs think, conceptualize their worlds, solve problems and apply their intelligence. With this knowledge, we can form even stronger bonds with our loyal canine companions.

The Role of Breed and Personality

When it comes to mental stimulation for dogs, it's important to consider how breed and personality impact their cognitive needs. A dog's breed can influence their energy level, trainability, and natural behaviors. Herding breeds like Collies and Shepherds tend to be highly intelligent and energetic, needing plenty of mental and physical exercise. Scent hounds such as Beagles have a strong prey drive and may be motivated by tracking games and puzzles involving smell. Terriers like Jack Russells are independent thinkers who benefit from training that challenges their problem-solving skills.

While general breed traits provide insight, each dog is an individual with their own distinctive personality. Just as people have different learning styles, some dogs are more visual while others respond better to reward-based training. Pay attention to your dog's unique quirks - the activities they engage with, their play style, how they interact with toys. This will give you clues into how they think and learn best. A dog that gets easily bored may need more frequent stimulation, while a anxious dog may thrive with lower-key games that build confidence.

There are key factors that influence a dog's cognitive development and abilities. Nutrition is crucial, as a poor diet can impair brain function, while high quality foods support neurological health. Physical activity also promotes circulation and oxygenation to the brain. Socialization and training during the first year are critical for developing mental capacity. Just like children, dogs have developmental milestones for acquiring certain skills.

Between 3-16 weeks old, puppies undergo rapid brain growth and are primed to learn basic commands, social rules, and how to problem solve through play. Exposing pups to novel sights, sounds and textures during this period can have lifelong benefits. Adolescence from 6 months to 2 years brings new training challenges as pups push boundaries and need mental stimulation to avoid bad habits. Mature adult dogs from 2-7 years benefit from learning complex tasks and commands to strengthen mental acuity. As dogs enter their senior years, mental exercise helps maintain cognitive function and slow age-related decline.

Tailoring activities based on your dog's stage of development will aid their mental growth. Puppies have shorter attention spans so 5-10 minute training sessions are most effective. Adolescent dogs need more significant mental and physical challenges like agility courses. Senior dogs may prefer slower paced exercise like easy hikes or food puzzles. Knowing your dog's personality andMatching exercises to your dog's age, breed traits and personality maximizes their engagement and interest in mental stimulation.

Observing your dog's daily habits provides insight into their mental state. Warning signs of boredom include pacing, unnecessary barking, destructive chewing, hyperactivity, or lethargy. Dogs experiencing anxiety may compulsively lick themselves, avoid contact, or be prone to separation distress. Seeking attention through whining or jumping up usually signifies a need for engagement. If your dog exhibits problematic behaviors, increasing mental stimulation is an effective solution.

Many undesirable habits stem from a dog having excess mental energy and no acceptable outlets. Teaching alternative behaviors through reward-based training redirects your dog's needs in a positive way. For example, if your dog jumps on visitors, teach them to fetch a toy instead to expend that energy. Increase the challenge by having them wait longer durations before fetching. Mental exercise also naturally reinforces desired behaviors by occupying your dog's mind. A tired dog from learning new commands will have less inclination to bark or get into trash cans.

Consistency and routine are important when implementing mental stimulation. Dogs thrive on predictability - having a set schedule for brain games, training and play times prevents boredom and behavior issues. Start each day with 5-10 minutes of training exercises focused on mastering a new skill or trick. Incorporate food puzzles and interactive toys into your dog's solo playtime - items like Kongs and treat balls provide mental challenges. Take your dog on an enriching afternoon adventure like a hike or trip to a new park once a week. Hold mini training sessions during downtime in the evenings. Adhering to this regimen keeps your dog's mind engaged and less prone to destructive behaviors.

Some dogs require more intensive intervention to truly transform problem behaviors. Maggie, a Jack Russell terrier, was surrendered to a shelter due to severe separation anxiety and destructive chewing whenever left alone. After adoption, her new owner discovered that increasing walks alone did not reduce Maggie's distress. She began leaving Maggie with food puzzles that required focus and effort to retrieve treats. This mental stimulation left Maggie relaxed and settled versus panicked when separated. Within two months, her anxiety and chewing had ceased with the implementation of regular mental exercise.

For Timmy, a hyperactive Australian Shepherd, his extreme excitability and tendency to jump on guests strained his relationship with his family. A trainer recommended teaching Timmy agility courses in the backyard, which required focus and gave him an approved activity for releasing energy. Daily agility practice curbed his uncontrolled behavior, and he learned to instead retrieve his favorite tug toy for guests to throw. Providing this intense mental and physical stimulation enabled Timmy's transformation into a composed, obedient dog.

The positive impacts of mental stimulation on canine behavior are clear. Adjusting activities based on your dog's needs and natural tendencies can work wonders for reducing problematic habits. But implementing brain games must be an enjoyable experience for you and your dog. Brain training should never feel like a chore - keep things lively by regularly introducing new games and challenges. Consistency is key, but allowing your dog choices gives them agency and heightens engagement. Track your dog's progress and celebrate breakthroughs, no matter how minor. With patience, positivity and persistence, mental exercise can shape your dog's behavior for the better while deepening the human-canine bond.

Cognitive Development in Dogs

A dog's mental capacities develop rapidly during the first months of life. Like humans, puppies go through critical developmental stages that shape their abilities. Understanding cognitive growth allows owners to provide age-appropriate stimulation. This primes puppies for training and establishes good behaviors.

The neonatal period, from birth to 2 weeks, focuses on basic physiological functions like eating and temperature regulation. Puppies also begin bonding with their mothers and littermates. Sensory development starts as eyes and ears begin working. These early weeks establish the foundation for good health. During the transition period from 2-4 weeks, puppies become more alert and mobile. Social skills develop through play and interactions with littermates. Puppies learn "bite inhibition" and other social cues.

At 3-16 weeks in the socialization period, puppies absorb key lifelong lessons. Positive exposure to people, environments, sounds, and handling sets them up for confidence. This primes their brains for communication and emotional growth. Fear responses also start emerging, making gentle introductions important. Puppies who miss socialization can develop anxiety or aggression later on. Owners should gradually introduce new stimuli while ensuring positive experiences.

The most intensive brain development happens from 6-24 weeks in the learning period. Puppies gain the cognitive maturity to understand training. This critical window is when owners should start obedience and manners lessons. Puppies pick up on patterns and cues fastest now. Exposing them to varied sights and sounds continues expanding mental horizons. Patience is important, as attention spans remain limited. Keep training sessions short and rewarding.

From 6-18 months, puppies transition into sexual maturity and adulthood. Physical growth slows down while mental maturity continues developing. Dogs in this adolescent period are primed to learn complex tasks. Their energy makes them highly responsive to training if owners channel it productively. More advanced commands build on the foundation established in the learning period. Sports like agility and flyball are ideal for adolescent dogs.

Though the early months are vitally important, cognitive enrichment should continue into adulthood. An adult dog's mental faculties remain flexible, allowing owners to teach new tricks and behaviors. Continuing games and social time also helps stave off age-related decline later on. Keeping adult dogs engaged and challenged prevents boredom or anxiety issues.

Senior dogs from around 8 years onward start experiencing canine cognitive dysfunction. This condition shares many similarities with Alzheimer's disease. Gradual neuron dysfunction affects memory, learning, awareness, and training responsiveness. Providing mental stimulation helps preserve existing brain pathways and function. Adapting exercises for limitations keeps senior dogs mentally active. Medications and supplements can also alleviate or prevent symptoms.

In summary, understanding the cognitive milestones in a dog's development provides guidance for owners. Tailoring mental enrichment to each stage stimulates growth and prevents issues. The early months build a foundation while adolescence and adulthood allow dogs to refine skills. Keeping senior dogs' minds active preserves function for as long as possible. Adjusting training and games to match each dog's abilities ensures they reach their mental potential at every life stage.

Recognizing Signs of Boredom and Anxiety in Dogs

Like humans, dogs experience complex emotions and mental states. Two common negative states that owners should look for are boredom and anxiety. Left unaddressed, these can lead to problematic behaviors and impact a dog's well-being. Recognizing the signs of boredom and anxiety allows owners to intervene early and take steps to improve their dog's mental health.

Boredom stems from inadequate mental stimulation and physical activity. Dogs require daily enrichment to engage their minds and fulfill their instinctual needs. Without outlets for their energy and intelligence, dogs easily become bored. This manifests in telltale behaviors. Destructiveness like chewing, digging or barking can indicate boredom. A dog may also pace, whine or appear restless when unoccupied. Attention-seeking behaviors like constantly bringing you toys or poking you to play suggest boredom.

Some additional signs include lack of energy, loss of interest in training or toys, and irritability. A bored dog's body language may appear mopy - tail and ears lowered, slow movements, yawning. They may watch the door for you to return. Boredom is often worse with under-stimulated breeds like huskies, herders and hunting dogs. Without sufficient activity tailored to their needs, they act out. Providing interactive toys, food puzzles, chews, and changes of scenery can relieve boredom.

Anxiety can stem from fear, insecurity, lack of socialization or stressful environments. An anxious dog struggles to relax and feels on edge. Behaviors like panting, pacing, whining, hiding, shaking and seeking constant owner contact can reflect anxiety. Loss of appetite, digestive issues and increased urination/defecation may also occur. Self-soothing behaviors like licking or chewing paws are common too.

Anxious body language includes a lowered or tucked tail, flattened ears, bunched facial muscles, avoiding eye contact, and a lowered body posture. Startling or reacting fearfully to noises, people or environments indicates anxiety. Separation anxiety when left alone is another manifestation. Triggers range from storms to car rides to handling by strangers. Genetics and traumatic experiences impact tendency toward anxiety.

To identify boredom or anxiety, look for changes in your dog's typical demeanor. Keeping a journal of their behaviors can help track any developments. Note times of day or scenarios when unwanted behaviors occur to detect patterns. Pay attention to your dog's eyes, ears, mouth and tail - their language reveals inner states. Subtle signs often precede more overt acting out, so early intervention is key.

If your dog exhibits multiple boredom/anxiety symptoms regularly, consult a vet to address medical causes. If the behaviors align with situational triggers, focus on modifying their environment and routine to meet their needs. Enrichment, training, socialization and bonding time can drastically improve boredom and anxiety. Anti-anxiety medications, pheromones and positive conditioning may also help for difficult cases or acute situations like thunderstorms or fireworks.

In summary, tuning into your dog's body language and mindset allows you to identify boredom and anxiety early. Implement prevention and training before behaviors become ingrained. A mentally stimulated, confident dog exhibits less distress and fewer problematic behaviors. Meeting their instinctual needs for activity while building trust and safety is key. With patience and compassion, you can help relieve your dog's boredom and anxiety and transform their state of mind.

CHAPTER 2

THE CONNECTION BETWEEN BEHAVIOR AND MENTAL STIMULATION

The Impact of Mental Exercise on Behavior

Mental stimulation is a critical component of a dog's overall health and wellbeing. When their minds are challenged through games, training and puzzle solving, the behavioral benefits are immense. Regular mental exercise enables dogs to channel their mental energy in a positive way, preventing boredom and destructive behaviors from taking root.

The inherent needs of dogs as intelligent, active animals means their behavior suffers without adequate mental engagement. Dogs thrive when they can put their problem-solving skills and instincts to use. Mental exercise taps into their natural drives - tracking scents, hunting prey, learning commands - while providing the mental "workout" they crave. Just as physical activity exercises a dog's body, mental stimulation keeps their minds fit and engaged.

Mental fatigue from constant stimulus without the chance to really exert themselves mentally leads to poor conduct. Bad habits like excessive barking, chewing, digging and hyperactivity can manifest when dogs have under-stimulated minds. Providing an outlet for their mental energy prevents frustration and restlessness. Brain games require focus and cognitive effort, leaving dogs feeling fulfilled rather than behaving destructively out of boredom.

Obedience training and learning new commands are forms of mental stimulation that curb unwanted actions. When dogs have to concentrate and focus to master a task, their likelihood of misbehaving from boredom is reduced. For example, a dog prone to jumping on visitors can instead be taught to go grab their toy on command when someone arrives. Training impulse control by rewarding patience and gradually increasing duration is also mentally tiring. A dog tasked with waiting before eating or going through doors is less inclined to react impulsively overall.

Mental challenges build confidence as well as obedience. Dogs feel a sense of accomplishment when they master an activity or trick through training. This boosts self-assurance while reinforcing the human-animal bond through positive reinforcement and play. Dogs taught to navigate obstacle courses or detect scents gain assurance in their abilities. This translates to better manners, social behavior, and responsiveness to owners.

Stimulating a dog's problem-solving skills curbs unwanted chewing and destruction. Puzzle toys that hide treats challenge dogs to manipulate objects to earn rewards. A dog focused on freeing kibble from a tricky toy has no interest in destroying household items. Rotating puzzle toys keeps canine minds engaged in positive play versus negative chewing habits. Interactive games that require mental strategy like hide-and-seek also reduce damage by providing an outlet.

Many dogs engage in destructive chewing simply to relieve stress and anxiety. Mental stimulation provides an alternate calming outlet for worried dogs. New experiences and environments encountered on daily outings offer mental distraction from stressors. Learning commands and training exercises boosts confidence while tiring the mind. Focusing their energy on thinking during activities leaves less room for anxiety. Providing outlets like chew toys and food puzzles also prevents boredom-related stress. A mentally fulfilled dog is calm and relaxed at home.

For insecure, reactive dogs, mental stimulation develops self-control and reinforces proper social responses. Activities that build focus, impulse control and obedience help ensure appropriate reactions to triggers like strangers or loud noises. Training an alternate behavior such as sitting to get a treat when encountering another dog helps prevent aggressive reactions born of fear and stress. Mental exercise is crucial for reactive dogs to correct unwanted behaviors.

Hyperactive dogs greatly benefit from intense mental stimulation tailored to their energy level. Breeds like Border Collies and Aussies thrive on agility training, advanced tricks and trials that challenge both mind and body. Developing their intelligence and abilities positively channels their hyper-ness into constructive pursuit versus chaotic conduct. Mentally vigorous exercise leaves them fulfilled and calm, reducing negative behaviors.

Providing adequate mental stimulation aligns with a dog's innate drives and needs. Rather than suppressing natural behaviors, it allows acceptable ways to satisfy them. Herding dogs can learn to round up toys instead of nipping people. Terriers inclined to dig get designated sandboxes. Retrievers obsessed with fetch play hide-and-seek with toys. Tailored brain games curb bad habits while enhancing a dog's quality of life and relationship with owners.

With the strong link between mental stimulation and improved behavior, implementing brain training is hugely impactful. Consistency is key - dogs need daily opportunities to expend mental energy through play, training and puzzle solving. Make mental exercise part of your dog's routine.

Try introducing new toys and rotating them to keep things interesting. Schedule training sessions every day working on mastering commands or tricks. Take your dog on outings like hikes that provide cognitive stimulation. Feed meals using puzzle toys or by training them to work for it through obedience commands. These actions prevent boredom while fulfilling your dog's needs.

Adjust intensity and type of games based on your dog's age, abilities and preference. A hyper young Lab may need complex puzzle toys and long play sessions, while a senior Pug prefers short training bursts and nosework games. Providing the right level of mental engagement prevents restlessness and misconduct. Always keep sessions positive using rewards and praise.

To get started enriching your dog's mindset and improving their conduct through mental stimulation:

- Get to know your dog's personality and natural instincts to tailor activities.

- Incorporate 5-10 minute training sessions daily to build focus and skills.

- Allow interactive playtime with food puzzles and toys to prevent boredom.

- Schedule outings that provide new sensory input and environments to explore.

- Rotate different brain games to keep their mind engaged.

- Adjust intensity and duration based on age, ability and energy levels.

- Reward and celebrate incremental progress to boost confidence.

With consistency, patience and an enthusiastic attitude, mental exercise can profoundly shape a dog's behavior for the better. A mentally stimulated dog is a happy, well-adjusted dog.

Identifying and Addressing Problem Behaviors

Problem behaviors in dogs often indicate underlying issues that need resolution. Identifying the root cause is the first step toward correcting negative habits. Common problems like aggression, anxiety, chewing, digging, or elimination inside stem from natural canine tendencies. While prevention is ideal, it is never too late to address challenges through compassionate training.

The motivation behind a dog's problematic behavior provides crucial insights. Aggression may reflect fear, pain, guarding instincts, or lack of socialization. Anxious dogs may have high emotional sensitivity, poor early handling, or negative past experiences. Excessive chewing or digging channels boredom or stress into destruction. Marking or eliminating inside signals incomplete house training or separation anxiety.

Once the underlying motivation is determined, owners can identify contributing factors. Lack of exercise, mental stimulation, or socialization are common issues. Changes in routine, moving, a new family member, or illness can also trigger problems. Dogs fed low quality diets may behave erratically. Medical causes like seizures, dementia, or pain must be ruled out. Pinpointing the source is the best starting point for meaningful change.

A veterinary exam can identify any health factors. Bloodwork, neurological testing, or behaviorist referrals provide insights. Patience, routine, positive reinforcement, and environmental management set dogs up for success. Removing triggers like access to shoes helps break destructive habits while training progresses. Confinement, crates, exercise pens, or baby gates limit access when owners are away. Preventing rehearsal and rewarding good behavior shapes it over time.

For anxiety, establishing a predictable routine and schedule can reduce uncertainty. Calming aids like pheromones and supplements lower stress. Counterconditioning shifts negative associations to positive ones using high value treats. Desensitization gradually increases exposure to triggers at a dog's comfort level. Learning cues for "settle" teaches dogs to relax on command. Confidence building through trick training empowers anxious dogs.

Aggressive behaviors require professional intervention beyond basic training. A certified applied animal behaviorist should conduct detailed evaluations for safety. Treatment may involve medication, management, desensitization, and counterconditioning. More serious aggression requires extensive modification and control. Rehoming is an option if quality of life remains poor despite best efforts.

Chewing or destructive issues respond well to more exercise, play, and chew-worthy toys. Stuffed frozen Kongs provide healthy chewing during alone time. Lick mats serve similar purposes. Rotating novel toys maintains interest. Teaching "leave it" and providing appropriate outlets focuses energy constructively.

House soiling requires revisiting potty training basics. Following a predictable schedule of bathroom breaks and praise strengthens bladder control. Restricting access until housetraining is solid prevents accidents. Thorough enzymatic cleaning eliminates odors. Examining factors like UTIs, new stressors, or incontinence ensures medical issues are not involved.

In many cases, prevention is the best solution. Early socialization, training, and meeting needs for exercise and enrichment head off future problems. Still, any dog can develop unwanted habits. Remaining patient, identifying motivations, and creating a plan tailored to the individual resolves many issues. Some behaviors may persist to varying degrees. But compassionate training focused on building skills, relationships, and rewards enhances quality of life.

Reinforcing Good Behaviors

Reinforcement strengthens desired behaviors using rewards and positive associations. By reinforcing actions owners want to see, dogs learn which habits earn benefits. Strategic reinforcement establishes reliable cues, commands, and manners. Understanding timing, motivation, and tailored rewards are key in training.

Reinforcing good behavior should happen immediately following the action. Even a few seconds delay can disconnect it from the reward. Using an event marker like a clicker precisely identifies the moment to reward. Food, toys, praise, or play incentivize repeating desired responses. Randomizing and limiting rewards prevents satiation. Keeping a variable schedule nourishes persistence.

Knowing what motivates each dog determines the best reinforcers. Food-driven dogs respond well to treats or kibble. Playful dogs value toy play or chasing games as rewards. Praise oriented dogs appreciate physical affection and verbal reinforcement. High value rewards like real meat maximize motivation until behaviors solidify. Avoid reinforcing with fear, intimidation, or punishment.

Shaping techniques gradually reinforce approximations towards a goal behavior. Reinforcing incremental steps cultivates complex skills like agility maneuvers or service tasks. Capturing naturally occurring actions that can be put on cue also utilizes shaping. Strategic reinforcement develops behaviors from scratch.

Maintaining reinforcement is key as behaviors strengthen. Transitioning to intermittent, randomized reinforcement on a variable schedule keeps dogs engaged without expecting constant rewards. Unpredictable rewards sustain behaviors better long-term than predictable patterns. Slowly increasing difficulty levels and criteria raises the bar while preventing frustration.

Common reinforcement mistakes include inconsistently rewarding behaviors, especially when first teaching them. Fading rewards too quickly can also undermine training. Reinforcing incorrect or unwanted actions accidentally is another common pitfall. Ensure timing precision and that only the intended behaviors receive rewards.

Prioritizing positive reinforcement over punishment or corrections is the most ethical and effective approach.rewarding incompatible alternative behaviors redirects dogs constructively. For example, reinforcing "sit" makes jumping up less rewarding. Reinforcing relaxed behaviors prevents anxiety and reactivity issues. Building new associations rewards emotional regulation.

Reinforcement principles extend beyond formal training to everyday behaviors. Dogs repeat actions that result in good consequences. Rewarding calmness and politeness reinforces manners and self-control. Providing outlets like chew toys reinforces using appropriate items. Reinforcing responsiveness to cues improves reliability off-leash.

Effective reinforcement requires reading subtle cues from dogs. Paying close attention helps owners time rewards precisely. Noticing trends in motivations allows customization. Observing body language identifies when to keep shaping versus add difficulty. Adapting reinforcement strategies based on daily energy optimizes progress.

In summary, intelligently utilizing reinforcement develops well-trained dogs who love cooperating. Rewarding generously in the early stages motivates persistence. Fading treats prevents bribery behaviors. Varying rewards maintains anticipation while avoiding satiation. Prioritizing humane, positive methods fosters mutual trust and understanding. Customized reinforcement tailored to each unique dog enhances communication, reliability, and the human-canine bond.

The Role of Routine and Consistency

Dogs thrive on routine and consistency. Implementing structure through regular schedules and training rituals taps into their natural appreciation of predictability. A consistent lifestyle and approach prevents confusion and aids obedience. Dogs learn best when expectations are clear and patterns repeated. Making good habits routine is key.

Establishing set times for activities like feeding, walking and playing creates a sense of order. Adhering closely to schedules in a dog's early training stages reinforces the significance of routines. Dogs are excellent time-keepers attuned to when events typically occur. Signaling walks, meals or playtime with cues like leashing up or preparing food at consistent times helps dogs anticipate and adapt to a routine.

Varying routines randomly can disrupt a dog's security. Gradual changes introduced incrementally are better tolerated once a baseline schedule is solidified. Routine also involves repetition - regularly reinforcing training, so commands become reflexive over time. Repeating games or walks along the same route taps into dogs' enjoyment of the familiar. Routines make dogs feel safe.

Household rules should also be well-defined and applied uniformly. Expectations around manners, sleeping locations or access to furniture must be consistent. Allowing behaviors sometimes but not others sends mixed signals. Maintaining consistency even when away from home prevents confusing rule-bending when traveling or visiting. Adhering to a training regimen builds good habits.

A key advantage of routines is reducing anxiety and undesirable behaviors. Structure minimizes stressors that trigger reactivity. Dogs prone to excitability, aggression or fearfulness benefit immensely from routines to increase stability. Military-style rigidity isn't necessary - aiming for general consistency reduces chaos. Plans may adjust as needed, but maintaining order and steady patterns benefits a dog's mindset.

Impulse control is also strengthened through set routines. Activities like sitting before exits or calmly waiting for meals to be served require obedience. Default behaviors rehearsed diligently as routine prevent problematic conduct. Routines allow dogs to anticipate situations and respond appropriately when those circumstances arise. Regularity also aids housebreaking through scheduled bathroom breaks.

While allowing flexibility, consistency provides benefits:

- Security from predictability and familiarity

- Reduced stress and anxiety, increased confidence

- Strengthened training through repetition

- Impulse control from rehearsed default behaviors

- Better household manners and obedience

- Easier troubleshooting of issues

Achieving consistency requires dedication, especially with challenging dogs. But structure and routine have calming effects on canine psychology. Dogs ease into lifestyles with natural rhythms and consistency. Make routines positive - regular playtime and training sessions, not just rigid rules. Inflexibility can have detrimental impacts too.

Balance is key - adaptable structure tailored to a dog's needs. Puppies require more rigidity initially, with mature dogs handling flexibility better. Active breeds need more exercise consistency than lower energy dogs. Overall, dogs thrive when routine provides security but allows for novelty and change too.

In summary, consistency through schedules, repetition and rules provides critical stability for dogs. Routines leverage their preference for predictability. Regularity reduces anxiety, strengthens training, encourages impulse control and promotes good conduct. Developing secure, positive routines adapted to a dog's unique needs unlocks their potential for obedience, confidence and better manners.

Case Studies: Transformations through Mental Exercise

Mental stimulation can profoundly impact a dog's behavior, as seen in these real-world case studies of canines transformed through brain games and training. Analyzing these examples provides insight into the power of mental exercise for modifying conduct and deepening the human-animal bond.

Lucy, a 2-year-old beagle, was surrendered to the local shelter due to severe separation anxiety and destructive behaviors. When left alone at home, she would bark excessively, eliminate indoors, and chew up furniture and shoes. The shelter staff determined Lucy's destructive habits stemmed from stress when separated from people.

To curb Lucy's anxiety, her new owner started with short departures while providing Lucy with puzzle toys filled with treats. Returning home to find the puzzles solved reinforced that being alone was not scary. Over several weeks, the duration Lucy was left alone gradually increased, always with distracting brain games. Within 2 months, Lucy could be left for 4 hours without incident, her separation anxiety conquered through mental stimulation.

Winston, a hyperactive 1-year-old Jack Russell terrier, jumped excessively on guests and could not calm down, trying his owners' patience. He was enrolled in a training class focused on agility courses and other games that required concentration and discipline. After 6 weeks of daily agility training using ramps, tunnels, and hoops, Winston's uncontrolled behavior diminished. He learned to wait politely for attention rather than jumping. Winston's energy was channeled productively through mental and physical exercise.

Shadow, a 5-year-old Labrador retriever, was well-trained but overweight and lethargic, concerning his owners. They implemented a new regimen of daily walks, portioned meals, and snuffle mats that forced Shadow to forage for kibble in fleece fabric. After 2 months of this mental and physical stimulation, Shadow lost 15 pounds and regained youthful energy. His focus and responsiveness also improved thanks to brain games and exercise.

Callie, an 8-month-old stray Husky, did not know any commands or manners when adopted. She jumped, mouthed, and was impossible to leash walk without pulling constantly. Callie's new family enrolled her in a 6-week basic obedience group class, where she thrived learning commands and agility skills. Daily 15-minute training sessions at home reinforced these lessons. Within 2 months of adopting Callie, she mastered leash walking, sit, stay, down, and leave it commands through dedicated mental work and repetition.

Diesel, a 3-year-old German shepherd dog, reacted fearfully to strangers entering his home, cowering behind his owner. A trainer created a customized desensitization plan for Diesel using positive reinforcement. Over 6 weeks, Diesel's owner rewarded calm behavior as strangers appeared at increasing distances, building Diesel's confidence. Soon he could retrieve toys for guests once fearful interactions, his wariness conquered through mental counterconditioning.

Riley, a 7-year-old mixed breed dog, began urinating when left alone after his owner returned to the office. The vet found no health issues, so his owner tried providing Riley with frozen Kongs when departing to distract and calm him. After a month Riley stopped relieving himself indoors, comforted by the mental stimulation of treat puzzles during alone time.

These examples demonstrate how targeted mental stimulation - from basic obedience, to agility training, to puzzle toys - can drastically improve a dog's conduct, confidence, and bonding with humans. While breeds, ages and needs differed, all these dogs benefited from having their minds challenged in healthy ways. With consistency and incremental training, the right mental exercise can rectify anxiety, boredom, excitability, fear, and stress. Riley, Winston, Diesel, Callie, Shadow and Lucy found improved purpose, temperament and connection through mental stimulation.

The key factors in the transformations included:

- Tailoring activities to each dog's needs and abilities. Lucy's separation anxiety was approached differently than Winston's hyperactivity.

- Implementing mental exercise incrementally and steadily, not rapidly overwhelming dogs. Shadow's regimen occurred over months.

- Making brain games and training positive experiences through rewards and praise. Callie thrived in her class through positive reinforcement.

- Ensuring the amount and intensity of mental stimulation matches the dog's capacities. Diesel's desensitization occurred gradually.

- Providing outlets for anxiety such as Kongs or training rather than scolding dogs. Distracting Riley with puzzles prevented accidents.

- Committing time and effort daily. All dogs improved with regular, quality mental engagement.

The proof is clear - mental exercise can profoundly impact dog behavior. From basic obedience to high-level training, brain stimulation gives dogs productive outlets for their energy and abilities. With an informed approach catered to your dog, mental learning yields impressive conduct breakthroughs and stronger human-animal bonds. Keep an open mind, celebrate small victories, and invest time into your dog's mental health for mutually rewarding results.

Laurel Marsh

CHAPTER 3

THE POWER OF POSITIVE REINFORCEMENT

Understanding Positive Reinforcement

Positive reinforcement strengthens behaviors by rewarding them. Applying rewards like food, toys, praise, or play makes desired actions more likely to recur. Positive reinforcement is the cornerstone of effective, ethical dog training. Understanding its nuances enhances communication and trust.

The basics involve providing positive reinforcers immediately after behaviors occur. This timing connection associates the reward with the action. Randomly delivering rewards on a variable schedule prevents expectations. Varied reinforcement encourages persistence better than predictable patterns. Generous early rewards prime motivation. Discontinuing reinforcement once learned preserves reliability.

Ideal positive reinforcers are things dogs naturally find rewarding. Food, freedom to explore, chasing prey, and social interaction tap into innate motivations. Customizing rewards based on individual preferences optimizes effectiveness. High value rewards like real meat up the ante for initial training. Using the lowest level rewards needed prevents bribery.

Reward delivery methods include hand feeding, praise, toys, and play. Food dispensed via clickers and target sticks pinpoint precise moments. Markers like "yes!" build associations. Integrating play and affection makes bonding part of training. Keeping reinforcement fun and engaging builds enthusiasm. Avoid physical or verbal corrections.

For best results, break down behaviors into incremental steps using shaping techniques. Reinforcing approximations toward a final goal lets dogs succeed. Fading food lures transitioned to hand signals illustrates this. Gradually raising criteria and difficulty level keeps challenges engaging. Impulse control games teach patience and focus.

Common mistakes include inconsistent delivery, poor timing, or limiting rewards too quickly. Reinforcing incorrect behaviors also undermines training. Ensure timing precision links rewards only to desired actions. Weaning off treats prevents dependency once behaviors are reliable. Patience prevents frustration while dogs figure cues out.

Positive reinforcement has many advantages over correction-based training. Rewards motivate better without suppressing behaviors. Dogs understand precisely which actions earn payoffs, avoiding confusion or fear. Bonding occurs through cooperation. Dogs remain optimistic and engaged with training. Risks of fallout like aggression are avoided.

Reinforcement principles extend beyond formal obedience. Rewarding calm, polite behavior in everyday life builds consistency. Providing appropriate outlets like chew toys reinforces constructive choices. Reinforcing responsiveness to cues solidifies reliability off-leash. Integrating training into play mixes fun with function.

The reinforcement approach is customizable for each unique dog. Paying close attention to an individual's preferences, motivations and temperament allows personalized methods. This insight comes from observation and relationship building. Custom-tailored training enhances communication, reliability and trust.

Positive reinforcement correlates with reduced stress and increased wellbeing. Motivation comes from within instead of outside pressure. Progress is achieved through incremental development focused on strengths. Deep understanding replaces control. Challenges become opportunities for growth. With patience and compassion, positive reinforcement transforms lives.

The Basics of Reward-Based Training

Reward-based training utilizes positive reinforcement to shape behaviors in dogs. Also known as positive training, it relies on giving rewards to encourage desired actions. This humane, effective approach harnesses your dog's innate desires to seek pleasure and avoid discomfort. Understanding the fundamentals of reward-based training establishes good relationships between dogs and humans.

The core idea is that behaviors rewarded will increase in frequency. So rewarding a dog with treats, praise, toys or play when they perform a desired action teaches them to repeat that behavior. The reward serves as positive reinforcement. This increases the likelihood of that action occurring again. Complex behaviors can be built by rewarding incremental steps toward the end goal.

Reward-based training contrasts with compulsion-based methods that apply punishment or force. Techniques like leash jerks, scolding and physical corrections teach dogs to avoid unpleasant stimuli. But this risks damaging trust and doesn't actively teach alternative good behaviors. Reward methods set dogs up for success by making the right choice enjoyable.

Proper timing is crucial when rewarding behaviors. The delivery of the reward must occur immediately after the desired response, within 1-2 seconds. This connects the behavior clearly with the positive outcome. Markers like clickers or verbal cues like "Yes!" bridge the precise moment of the behavior to the reward.

Consistency also matters. Reinforce every desired behavior at first, then move to intermittent reinforcement by rewarding only some instances. This strengthens the behavior over time. Keeping rewards variable prevents dogs from expecting a "pay-off" every single time. Phase rewards out gradually as behaviors become reliable habits.

Effective reward-based training employs these guidelines:

- Use enticing, high-value rewards like real meat, cheese or toy play. Avoid low-value treats.

- Get your dog's attention and focus before asking for a behavior.

- Give clear, concise verbal or visual commands.

- Deliver rewards quickly after the instant the dog complies.

- Mark correct behaviors immediately with clickers or words like "Yes!".

- Reward every repetition initially, then randomly provide rewards intermittently.

- Gradually phase out food rewards as behaviors improve.

- Use real-life rewards when possible - throw a ball to reinforce "fetch".

Common beginner mistakes include poor timing, low-value rewards, lack of consistency, and not requiring focus during training sessions. Patience is also needed, as dogs may not initially understand what earns rewards. Refine technique and motivate your dog by making training fun! Short, positive sessions are best for engagement and progress.

To begin reward-based training:

- Start with basic behaviors like sit, down, stay, come, heel.

- Break commands into small, achievable steps for rewards.

- Use high-value incentives tailored to your individual dog.

- Time delivery of rewards properly after desired response.

- Mark correct behaviors with clicker or verbal cue.

- Reward consistently at first, then intermittently.

- Keep sessions brief and fun - 5 minutes for puppies, 10-15 minutes for adults.

The benefits of reward-based training extend beyond teaching commands. This approach strengthens the human-canine bond through mutual motivation. Dogs learn to look to their owners for guidance and rewards. Positive reinforcement also boosts confidence in shy or fearful dogs. Reward-based training creates eager, attentive pets who partner happily with their families.

Common Mistakes and How to Avoid Them

Implementing positive reinforcement training with your dog can utterly transform their behavior and deepen your bond when done correctly. However, there are also common mistakes that can hinder progress. Being aware of these pitfalls and how to avoid them sets you up for training success.

A major mistake is misunderstanding positive reinforcement concepts. Rewarding a dog at the wrong time, like after they have stopped exhibiting the desired behavior, teaches nothing. The reward must come during or immediately after the wanted action to reinforce it. Constantly treating a dog without requiring effort or obedience does not constitute positive reinforcement either - the reward must correlate to a specific behavior.

Another critical error is relying on punishment rather than rewards. Techniques like leash jerks, scolding or startling a dog are ineffective for positive training. You cannot reinforce good behavior while scaring or causing your dog discomfort - this sends mixed signals. Only use rewards and withhold rewards to shape conduct. If your dog is receiving more punishments than treats in training, you need to reset.

Failing to properly motivate your dog is another common downfall. You must identify rewards with enough value to motivate them to focus and work. Food treats are reliable motivators for most dogs - experiment to find which ones your dog loves. Alternatively, some dogs are toy motivated - discovering objects your dog obsessively pursues can be leveraged for training. Always train when your dog is eager and engaged, not distracted or stressed.

Presenting training sessions that are too long or complex is a frequent issue. Young puppies have short attention spans and elderly dogs tire easily as well. Keep most training sessions to 5-10 minutes for average adult dogs, and end on a positive note. Gradually increase difficulty and duration at your dog's pace. Move through steps slowly, especially with complex behaviors - only proceed when your dog has mastered the last step fully.

Being vague or inconsistent with cues and commands leads to confusion and stalled progress. Use consistent verbal and hand signals each time you give a command. Avoid repeating or saying cues multiple times, which teaches your dog to ignore you until you beg. Give clear, precise signals in the same way every time to maximize your dog's comprehension.

Failing to "capture" and reward spontaneous good behaviors is another missed opportunity. If your dog voluntarily sits, lays down, comes to you, etc. without being cued, reward them! Capturing reinforces natural conduct you want to encourage. Missing these moments prevents positive behaviors from becoming habits. Have treats on hand at all times for potential capturing.

Not varying rewards can also cause dogs to lose motivation over time. Change up food rewards with different flavors and textures. Alternate between food treats, praise, toys and life rewards like getting to go outside or play with another dog. Using unpredictable rewards heightens your dog's engagement and interest in training.

Finally, a major but common error is accidentally rewarding unwanted behaviors. If your dog jumps on you and you push them off while saying "No! Down!" you just reinforced jumping up through attention. Only reward calm conduct, not exuberant behavior. Be mindful that unwanted responses are not inadvertently getting reinforcement.

There are simple ways to avoid these common pitfalls:

- Clearly understand positive reinforcement principles before beginning training.

- Never use punishment - only positive rewards and withholding rewards.

- Identify your dog's strongest motivators through testing different options.

- Keep training sessions short, regulated, and tailored to your dog's abilities.

- Use highly consistent verbal and hand signals for cues.

- Reward spontaneous good behavior whenever it occurs through capturing.

- Provide varied rewards - mix up food, toys, praise and life rewards.

- Be vigilant in only reinforcing desired conduct, not unwanted behaviors.

While training mistakes are inevitable, being aware of common errors will help you self-correct. Focus on rewarding the behaviors you want to see more of, while preventing reinforcement of conduct you want extinguished. Stay positive - progress may be slow, but you are laying a foundation for lifelong learning together. With dedication and patience, avoidable mistakes will fade away.

Building on Success: Increasing Difficulty Gradually

Growing skills requires strategic progression. As dogs master basic behaviors, raising criteria incrementally avoids frustration. Keeping challenges engaging without being overwhelming sustains motivation. Gradually introducing distractions and longer duration builds real world reliability. Planning small steps empowers success.

The early stages focus on reinforcement and creating positive associations. Keeping training sessions short with abundant rewards primes dogs to initial cues. Once consistently responding, begin requesting behaviors before delivering rewards. Randomly reinforce successful responses while gradually expect more effort.

Increasing criteria means requiring greater precision, speed, or duration. Only raise expectations after multiple successes at the current level. Asking for multiple repetitions before rewarding increases endurance. Slowly extend duration of sustained behaviors like "stays". Adding slight distractions like distance or minor noises builds real world proficiency.

Varying difficulty prevents boredom once basics are reliable. Balance easy repetitions to boost confidence with progressively harder challenges. Maintain a high rate of reinforcement as challenges increase to buoy motivation. Breaking down complex behaviors into smaller steps lets dogs progress. Fading food lures transitions cues to hand signals.

Shaping techniques reinforce successive approximations toward a goal.Rewarding slight improvements incrementally develops new skills. Breaking down long durations into intervals marks progress. Changing locations routinely boosts generalization. Introduce distractions methodically until maintaining focus is second nature. Raise criteria in small increments tailored to each dog's abilities.

Patience is essential as difficulty increases. Expect setbacks and plan repetitions to solidify skills. Increase challenge levels conservatively to maintain momentum. Allow extra time for proofing until behaviors are rock solid. Better to proceed gradually than overwhelm dogs and undermine progress.

Avoid common mistakes like rushing forward or insufficient repetition at each level. Ensure behaviors are truly reliable before adding new challenges. Watch for signs of frustration and revert to easier efforts to rebuild confidence. Repeating steps is better than moving ahead prematurely.

Increasing difficulty strategically lets dogs scaffold their abilities. Each step strengthens neural pathways, coordination, and understanding. Physically and mentally exercising dogs prevents boredom while deepening skills. Their capacity for focus and complexity expands through gradually raising the bar.

Adapting difficulty levels to suit each dog's abilities and temperament optimizes development. Paying close attention to body language identifies suitable pacing. Highly motivated dogs can advance more rapidly. Slower progression benefits shy or sensitive dogs. Customized training taps full potential.

Progress happens in incremental stages, not giant leaps. Patience paired with observant adjustments keeps moving forward without overwhelming. Celebrating small wins maintains optimism. Gradual but steady difficulty increase empowers dogs to believe in their growing abilities.

Case Studies: Positive Reinforcement in Action

Seeing positive reinforcement training in action illuminates how impactful reward-based techniques can be. Let's look at real-world examples and case studies showcasing the power of this approach.

Ginger, a 1-year-old Labrador Retriever rescue dog was extremely fearful of strangers and unfamiliar environments. She would bark, back away, and hide when anyone approached her. Ginger's owners wanted to improve her confidence and teach her basic obedience. They enlisted a positive reinforcement trainer who used high-value meat treats to reward any small instances of Ginger voluntarily moving closer to people or tolerating petting.

Within a few weeks, Ginger's body language relaxed as she associated strangers with delicious treats. Her compliance with basic commands like "sit" and "stay" also improved dramatically when each repetition was rewarded with praise and treats. Ginger became more outgoing and sociable using only positive techniques tailored to her fearful personality.

Simon was a young Belgian Malinois with limitless energy and no manners. He constantly jumped up, mouthed hands, and needed to learn impulse control. Simon's owners had tried scolding and physically correcting him but saw little progress. A reward-based trainer showed them how teaching incompatible behaviors could crowd out Simon's unwanted habits.

Asking for a "sit" before giving attention rewarded calm behavior and reduced jumping. Giving chew toys instead of hands for mouthing rewarded acceptable outlets for biting impulse. Simon learned over time that good manners earned him rewards. His exuberance was gradually channeled into training exercises focused on control.

Luna, a nervous Chihuahua, was reactive toward dogs on walks, lunging and barking. Her owners found walking her to be stressful. A trainer helped them implement positive reinforcement by coaching Luna to focus on them whenever they saw another dog. Rewarding eye contact and attention made Luna seek treats rather than fixating on triggers. Over several weeks, her reactivity diminished as her owners positively reinforced calm behavior.

In another case, Maya the elderly Greyhound learned "shake" using lures and rewards to strengthen her coordination and cognition. And rescued puppy Charlie overcame mouthing and nipping through positive redirection to chew toys.

These examples showcase how reward-based training transforms dogs' behaviors. The key principles used successfully include:

- Tailoring rewards - high-value, "jackpot" rewards for significant behaviors; frequent small rewards for repetition.

- Addressing motivations - use rewards that align with reasons for the behavior.

- Starting small - reward even tiny steps toward end goal at first.

- Reinforcing incompatible behaviors - teach alternate wanted behaviors.

- Building on success - incrementally shape complex behaviors through progressive rewards.

- Removing rewards for unwanted behaviors - don't reinforce bad habits.

- Staying positive - never scold; redirect to what you want.

While time-intensive initially, reward-based training creates long-term improvements. Resolving issues like fear, anxiety, hyperactivity and aggression positively is more effective and humane than punishment. Real cases prove reward-based training's ability to promote obedience, build confidence, increase sociability and strengthen the human-canine bond.

Laurel Marsh

CHAPTER 4

BRAIN GAMES FOR DOGS

The Principles of Brain Games

Brain games are an excellent way to provide dogs with mental enrichment and stimulation. But to reap the full benefits, it's important to understand the core principles behind choosing and implementing effective brain games. Following these key guidelines will lead to engagement, learning, and valuable strengthening of the human-canine bond.

The first principle is to select activities suited to your individual dog based on their breed traits, personality, and natural instincts. For example, brain games for a high-energy herding breed like a Border Collie should involve more vigorous physical and mental challenges than those for a smaller or low-energy breed. Play hide and seek with a scent-driven Beagle using treats to fuel their nosework. Engage a trainable Poodle with learning new tricks and commands to challenge their intelligence. Lean into your dog's innate inclinations when choosing brain games.

Secondly, brain games should align with your dog's current abilities, development stage, and progress. Start with basic games for puppies with shorter attention spans, then increase to more complex tasks as they mature and master fundamentals. Brain games for senior dogs may need adapting to avoid overexertion. Keep training sessions short and positively reinforced. Adjust games and difficulty as your dog's cognitive abilities grow.

Maintaining your dog's optimal motivation and engagement is also key. Incorporate their favorite treats, toys and types of play into games. Maintaining their interest and participation throughout prevents boredom. End each session on a high note, not when your dog loses focus. Regularly switch up and introduce new games to prevent monotony. Keeping your dog actively excited during gameplay yields the best mental stimulation benefits.

Additionally, dogs learn best through clear instruction and positive reinforcement. Use consistent verbal and hand signals when introducing new brain games and commands. Reward desired behaviors immediately and generously. Avoid scolding mistakes - just refocus your dog's attention and reinforce the right response. Keep a lighthearted, upbeat attitude rather than getting frustrated. Your positivity will elevate your dog's eagerness to participate.

Providing brain games on a consistent schedule is also fundamental. Dogs thrive on routine. Set up recurring game sessions at the same times daily. Integrate them into your dog's life, not just doing them sporadically. Daily mental engagement through schedules brain games prevents restlessness and boredom. Sessions can be brief but should be consistent.

It's also important to incorporate brain games into solo playtime. Rotating puzzle feeders, hide-and-seek toys, chew toys and more enables ongoing mental stimulation when you cannot directly engage your dog.Choose enriching toys that require manipulation, like food puzzles. Avoid leaving toys out constantly - rotate a few at a time to keep novelty and interest. Schedule interactive play when leaving dogs alone.

Additionally, brain games should take place in new environments to provide mental novelty. Take games on the road to new parks, trails, and locations. Exposure to new settings, sights and sounds sharpens mental acuity. Let your dog explore novel scenarios safely during brain games. This mental enrichment keeps them engaged with the activities.

Finally, brain games should be fun! Avoid drills or repetitive actions. Incorporate movement, exploration, bonding and praise into activities. Let your dog take the lead sometimes in choosing games and directing play. Make training feel like play, not work. Keep things exciting for you and your dog. Shared enjoyment strengthens your bond while optimizing mental benefits.

Following these brain game principles will maximize mental stimulation for your dog:

- Tailor games to your dog's breed, personality and instincts.

- Adjust difficulty as your dog's cognitive abilities develop.

- Maintain high engagement and motivation with exciting variables.

- Use positive reinforcement and clear instruction during games.

- Schedule consistent daily brain game sessions.

- Incorporate independent playtime with enriching toys.

- Provide games in new environments for mental novelty.

- Ensure gameplay is fun and enjoyable for both of you!

Adhering to these guidelines will result in brain games your dog loves. They will look forward to these mentally stimulating activities while you notice their improved focus and behavior. Braingames are a win-win for strengthening the human-canine bond!

Indoor Mental Exercises for Rainy Days

Inclement weather need not derail mental enrichment for dogs. Numerous activities provide cognitive challenges from the comfort of home. Creativity and environmental management transform rainy days into engaging brain games.

Food puzzle toys offer problem solving indoor play. Kibble-dispensing balls and tubes require manipulation to earn meals. Obstacle courses made from household objects build navigational skills. Hide and seek games tap into natural foraging instincts. Scentwork develops nose skills searching for hidden treats.

Indoors training sessions provide mental exercise along with reinforcing behaviors. Practicing obedience basics or tricks in new locations challenges dogs. Balance training with play breaks to maintain engagement. Vary commands and difficulty level to keep dogs thinking.

Physical and mental exercise can be combined by clearing space for indoor activity. Set up tunnels or low jumps in basements or garages for agility practice. Playing fetch down stairs or hallways adds exercise. Treadmill walking provides controlled cardio.

Adapting outdoor games for inside keeps dogs engaged. Indoor nosework hides treats to sniff out around the house. Hide and seek develops search skills finding family members. Learning new routes and paths rooms builds spatial awareness.

Providing stimulating chew toys alleviates boredom. Frozen Kongs with kibble and treats provide mental and physical chewing satisfaction. Bully sticks, horns, and tendons keep jaws busy. Rotating novel safe chews maintains interest. Supervise chew time to prevent destruction or consumption of household objects.

Reinforcing relaxation skills is key for high energy dogs. "Settle" and "go to mat" cues teach calmness on command. Providing puzzle toys stuffed with food focuses pent-up energy. Crating with a stuffed chew toy redirects restlessness.

Adapt games based on a dog's abilities and temperament. Lower energy dogs benefit from mellower activities like nosework and light training. Higher drive dogs need more vigorous play and food puzzles to prevent destructive behaviors. Monitor stress signals to avoid overstimulation.

Mental enrichment should engage a dog's natural instincts in a constructive way. Games and toys that tap into foraging, chasing, chewing, and problem solving provide an outlet. Following their lead on preferred activities keeps dogs invested. Frequent but shorter sessions prevent boredom from repetition.

Inclement weather teaches creativity and flexibility. Adjusting routines maintains enrichment amidst cabin fever. Indoor activities reinforce the human-animal bond through playful interaction. Mental exercise provides meaningful engagement, meeting canine needs for more than just physical activity. With innovation and dedication, dreary days enhance rather than detract from quality of life.

Outdoor Games for Stimulating the Canine Mind

Mental exercise is as important as physical activity for dogs. Engaging a dog's mind through games and puzzles stimulates them both physically and intellectually. Outdoor play provides fun opportunities to challenge your dog. Tailoring activities to your dog's abilities and drives takes their enrichment outdoors.

Nose work games leverage a dog's keen sense of smell for mental exercise. "Find it" starts simply by having your dog sit-stay while you hide treats in plain sight outdoors. Say "find it!" releasing your dog to seek the reward. Increase difficulty by hiding treats in bushes, trees or buried slightly underground. Advance to using favorite toys or articles of your clothing as scent articles for them to locate.

Another nose game is scattering small treats in grass or bushes and timing your dog as they hunt to find them all. Or place numbered containers holding different scented essential oils outdoors and train your dog to find specific scents on cue. These sniffing challenges engage your dog's brain.

For working breeds, setting up backyard agility obstacles like tunnels, jumps and weave poles allows them to practice athletic skills daily. Position several obstacles together and guide your dog through sequences, rewarding with toys or treats. Start slowly, ensuring safety and mastery on individual obstacles first. Vary patterns to keep your dog engaged.

Food-dispensing toys offer mental exercise too. Kongs stuffed with frozen wet food or puzzle toys containing kibble provide motivation to manipulate the devices for edible rewards. Allowing supervised play with these useful distractions exercises their brainpower. Rotate novel toys to prevent boredom.

For fetch-driven dogs, use ball launchers that throw balls far distances. The physical running and mental focus on watching the arc of the ball provides energy outlets. Or put old socks on your dog's paws and play chase - the slippery socks impede running slightly, creating a fun challenge.

Pups also enjoy "popping bubble wrap" outdoors. Let them zig-zag across bubble wrap taped to a yard surface, rewarding pops with kibble. This combines physical activity with problem-solving smarts to reach the objective.

Playing "find the handler" also stimulates your dog. Have a family member hold your dog while you hide behind a tree or bush. Call your dog's name so he tracks you down by your voice. Switch roles so your dog learns to stay until called to "find" you. This tests communication, obedience and focus.

Increase the challenge by hiding a favorite toy instead of yourself. Shuffle several toys while your dog waits, then give the cue to seek the hidden one. Start easy by completely visible hides before progressing to outdoor concealment. This taps into your dog's memory and smell.

Mental exercise should involve novelty and progressively harder tasks to stay stimulating. Maintain your dog's engagement through positive reinforcement. Short 5-10 minute sessions prevent mental fatigue. Monitor your dog closely during new games - don't progress too quickly or force participation. Keeping activities enjoyable prevents stress.

Discover creative, interactive games that test your dog mentally. Nose work, agility, food puzzles, recall challenges and hunting games all provide purposeful work for your dog's brain and instincts. Meeting their mental needs through outdoor play reduces unwanted behaviors. Regular brain games keep your dog sharp, enriched and engaged with the world around them.

Interactive Toys and Puzzles: A World of Options

Puzzle toys and interactive games provide dogs with invaluable mental enrichment and stimulation. The market offers an exciting, ever-expanding array of brain-building toy options to keep your dog happily engaged. With so many choices, what are the best interactive puzzles and games to stimulate your canine companion?

Food dispensing toys are a top pick for mental stimulation. Toys like the Kong involve your dog manipulating and moving the toy with their paws and mouth to dislodge kibble or treats. The effort of shaking and rolling the toy to release food challenges your dog's problem-solving skills. Models like the Kong Wobbler and Trixie Activity Flip Board have adjustable difficulty levels to increase the cognitive demand.

Another excellent interactive toy category is treat-release puzzles. These toys require your dog to slide, lift or shift puzzle pieces to uncover hidden kibble. Popular versions like the Nina Ottosson line or Trixie Activity toys have progressively harder levels to challenge your pup. Getting rewards out of intricate concealed areas engages your dog's instincts to forage and hunt.

Scenting and nosework toys are also fantastic brain builders. Snuffle mats and hides-a-squirrel plush toys encourage dogs to use their nose to root out kibble stuffed inside. Dispensing toys infused with your dog's favorite aromas add olfactory appeal. Activating your dog's powerful sense of smell is both tiring and exhilarating.

Toys that encourage fitness provide physical and mental exertion. CleverPet Hub is an interactive game system with touchscreen puzzles. Your dog learns to touch lighted targets on the screen to earn treats. This combines training, problem-solving and activity. Other toys like iFetch throwers turn retrieving into a game of skill.

Chews like frozen stuffed Kongs occupy dogs mentally and physically. Working to access part of the treat while cold numbs pain from sore gums. Long-lasting chews help prevent boredom and anxiety when you are away. Edible chews add incentive and appeal beyond standard toys.

Indestructible toys that can be stuffed with food or treats are another option for active manipulation. Brands like Goughnuts, West Paw and Planet Dog make tough toys dogs can safely bite, chew, roll and play with for hours. The mental work of accessing stuffed contents prevents destructive boredom. Durable enough for aggressive chewers, these toys last.

Interactive toys that provide cognitive challenges are great investments. More complex toys encourage mental maturity as your dog learns to solve increasingly difficult puzzles. Automated toys allow you to program timed release of food or adjust difficulty settings remotely. Smart toys provide a lifelike level of interactivity and variability.

Outdoor toys like treat-dispensing balls and puzzle boxes also encourage mental and physical exertion. Food puzzles buried in sandboxes or hidden around the yard motivate natural foraging instincts. Bringing interactive play outdoors adds variety and environmental novelty.

Beyond manufactured toys, many household items easily become brain games. Paper towel rolls stuffed with kibble, an empty plastic bottle for batting around, old socks tied in knots with treats inside - everyday objects encourage inventive play. Rotate novel items to stimulate curiosity.

The key is to incorporate toys that align with your dog's abilities, energy and preferences. Monitor your dog's interest level with new toys. Phase out items they lose interest in and introduce new puzzles. Provide a range of easy, moderate and difficult toys to challenge your pup. Rotate frequently to prevent boredom. With so many options, you can curate the ideal assortment of toys tailored to your dog's needs and instincts.

When selecting toys:

- Choose durable toys that match your dog's size and chew strength.

- Look for adjustable difficulty levels to keep your dog challenged as they develop mentally.

- Appeal to your dog's senses with scenting, sounds and textures to increase engagement.

- Include some toys that encourage physical activity, not just mental skills.

- Stimulate natural behaviors like foraging, hunting and chewing with appropriate toys.

- Refresh toy selections routinely to keep your dog interested in playtime.

With the amazing range of interactive puzzle and brain games now available, the possibilities for mental enrichment are endless. Whether you prefer technological gizmos, stimulating snuffle mats, crunchy chews, or basic household items - there are limitless options. Puzzle toys encourage vital cognitive growth and skills. Engage your dog's mind each day with their favorite brain-building activities for a mentally balanced pup!

Creating Your Own Brain Games

Tapping into owner creativity expands mental enrichment possibilities. Homemade toys and activities allow customization for each dog's abilities and preferences. Simple, low-cost brain games exercise neuroplasticity through problem solving skills.

Food puzzle toys provide interactive feeding. Simply drilled wood blocks or plastic bottles filled with kibble require manipulation to earn meals. Customizing hole size and difficulty adjusts challenge levels. Other materials like cardboard tubes, muffin tins, or pool noodles can be adapted into food puzzles.

Nosework develops natural scenting ability. Hiding treats around the house or yard creates search games. Use higher value rewards to motivate hunting. Start with easy finds, slowly increasing difficulty by obscuring treats more. Essential oil scents add complexity for advanced nosework.

Obstacle courses build physical and mental stamina. Arrange household objects like broom sticks, buckets, chairs into navigational challenges. Change patterns regularly so dogs must problem solve fresh layouts. Low platforms, ramps and tunnels build coordination. Caution using unstable or hazardous items.

Learning new skills and tricks keeps dogs engaged with owners. Keep a list of fun new tricks to cross-train. Shaping techniques reward incremental progress. Creativity expands possibilities - teach weaving through legs, rolling over, or catching treats in the air.

Hide and seek utilizes toy drive for mental exercise. Have dogs wait in another room while hiding a favorite toy. Trade off finding each other and retrieve toys. Increase difficulty by choosing harder locations over time.

Matching games exercise cognitive skills. Line up identical toys or objects of different colors/textures. Ask dogs to match two of the same item. Advance to matching multiple item pairs and identifying mismatched ones.

Sensory stimulation engages the mind through sound, scent, touch and sight. Pack textured toys, novel scents or recorded sounds into a box for dogs to explore. Rotate items to keep the experience new and interesting. Supervise to avoid ingestion.

Mental enrichment activities should align with a dog's natural abilities and instincts. Observe motivations and preferences to determine which DIY games will provide fulfilling challenges. Routine becomes boring - change games regularly to keep their minds flexible.

Homemade brain games allow customization for each dog's interests, energy level and abilities. Simple tweaks adjust difficulty to keep dogs engaged. Unleashing creativity enhances the human-animal bond through mutual play and problem solving. Quality time spent building and playing games strengthens relationships.

Laurel Marsh

CHAPTER 5

ADVANCED TRAINING TECHNIQUES

The Basics of Clicker Training

Clicker training employs positive reinforcement by using a clicker tool to precisely mark desired behaviors. The click sound becomes a conditioned reinforcer when paired with rewards consistently. This allows trainers to "capture" wanted behaviors at exact moments. Clicker training communicates effectively in your dog's language.

Clickers provide split-second timing to pinpoint behaviors to reward. The sound is unique and distinct from verbal praise. Dogs learn that click = treat. The consistency develops automatic reinforcement. Clicker training works by:

- Dog performs behavior

- Trainer clicks during behavior

- Dog receives reward

- Behavior strengthens due to consistent click-treat association

This timing precision is especially useful for shaping complex actions or teaching during distractions. The clicker bridges the instant of behavior to the reward delivery, preventing delay that can confuse dogs.

Clicker training focuses on reinforcing voluntary positive behaviors, not punishing unwanted ones. Strategic rewards teach the dog what you desire them TO do rather than what NOT to do. The clicker allows capturing great behaviors that happen spontaneously. You can then put them on cue with a verbal or hand signal.

Proper technique is vital. Use short click bursts rapidly - don't hold it down. Click only once to mark precise moments. Give rewards within 1-2 seconds of clicking. Vary rewards with food, toys and praise to prevent habituation. Wean off the clicker gradually as the behavior becomes learned.

Follow these guidelines when introducing the clicker:

- Charge the clicker by clicking then immediately rewarding with food several times.

- Time clicks for natural or lured behaviors at first. Move to behaviors you cue intentionally.

- Only click for the exact behavior you want to reinforce - don't over-click.

- Give the reward after the click, not simultaneously.

- Use high-value rewards in the beginning, then vary reinforcers.

- Keep initial sessions short and engaging until your dog understands the pattern.

Common mistakes include clicking too late, clicking for multiple behaviors simultaneously, delaying the reward after clicking, and failing to "charge" the clicker meaningfully. Be patient - dogs need repetition to grasp clicker training.

To train a new behavior with a clicker:

- Give a hand signal or lure to prompt the physical response.

- The instant the dog does the behavior, click then reward.

- Repeat steps 1-2 consistently until the dog reliably performs the behavior.

- Add the verbal cue immediately before the hand signal.

- Gradually phase out the hand signal as the dog responds to the verbal cue.

- Move to intermittent reinforcement by only rewarding some repetitions.

Examples of behaviors to train with a clicker include:

- Basic commands like sit, down, stay, come, heel.

- Addressing behavior issues like jumping up or barking.

- Complex skills like leg weaves or rolling over.

- Building confidence and engagement through toys or tricks.

Clicker training optimizes positive reinforcement, fosters enjoyment of training, and strengthens the human-canine bond. The clicker's precision focuses behaviors while the rewards motivate performance. Combining clear communication with consistency and patience leads to measurable improvements.

Whether addressing existing problems or teaching advanced skills, clicker training builds relationships of trust. Adopting positive methods creates enthusiastic dogs who look forward to interacting and learning with you.

Using Target Sticks and Lure Sticks

Target sticks and lure sticks are valuable tools for motivating and guiding dog behaviors. Proper technique strengthens communication and precision in training. Understanding the nuances of implementing sticks expands training possibilities.

Target sticks use targeting to direct focus and movement. Dogs touch their nose to the end of the stick on cue for reinforcement. This builds coordination and handling skills. Varying target stick positions guides precise movement through obstacles or patterns.

Lure sticks employ food or toys to coax behaviors before putting them on cue. Slowly fading lures transitions to visual hand signals. Lure sticks focus attention and demonstrate desired responses. Gently guiding the dog's head builds muscle memory until the behavior is mastered.

Stick length allows comfortable handling. Target sticks around 18 inches give dogs room to target without unwieldy length. Lure sticks can be shorter, between 6-12 inches, for more controlled luring. Lightweight dowels or whip sticks enable easy maneuvering.

Positive reinforcement methods make sticks highly effective. Immediately reward touching target sticks or properly following lure sticks. Food, toys, clicks, and praise incentivize repetition. Keep sessions short and engaging to build enthusiasm. Introduce sticks gradually through shaping and building value.

Target stick exercises develop coordination, impulse control and bonding. Cue targeting in different positions - ground level, waist height, overhead. Move the target unpredictably to sharpen reflexes. Send dogs through obstacles targeting the stick on the opposite side. Reward composure waiting to be cued.

Lure sticks communicate precise positioning for skills like heel, spin, crawl. Reward smallest attempts, gradually shaping and fading off food lures. Transition to hand signals once the behavior is learned. Alternate luring and reinforcing known behaviors to maintain reliability.

Avoid frustration by working in short sessions with liberal rewards. Read body language for fatigue or confusion and take breaks. Ensure basic targeting skills are solid before complex sequences or novelty. Prevent chasing or biting sticks by reinforcing calm control.

Using sticks builds comprehension, accuracy and responsiveness through clear visual cues. They channel energy into productive activities. Combining sticks expands possibilities - targeting stick positions can guide lure stick sessions. Customizing stick use for each unique dog optimizes communication.

With patience and creativity, sticks unlock training potential beyond basic obedience. They forge attentive partnerships built on trust and positive reinforcement. Advancing carefully while making training fun prevents drudgery. Target and lure sticks enrich skills, confidence and the human-canine bond.

Introducing Agility Training at Home

Agility training builds confidence, focus and athletic coordination in dogs. Running through obstacle courses provides physical and mental exercise. While professional classes are ideal, introducing agility foundation skills at home allows dogs to learn at their own pace. Home training reinforces key lessons and primes dogs for public training.

Select obstacles suitable for home use. Portable jumps with adjustable heights are easy to set up in yards or inside. Tunnels made of flexible hoops allow teaching the "tunnel" cue. Weave poles can be DIY'ed from PVC pipes and ground stakes. Always ensure safety - avoid hard surfaces, secure equipment and monitor closely.

Work on one piece at a time, in short sessions. Hand targeting teaches dogs to touch their nose to your hand. Hold your hand slightly above your dog's nose, say "touch" and mark/reward nose contact. Fade the hand lure and add the verbal cue "touch". Targets help guide dogs properly through obstacles.

Shaping a "go around" cue teaches circling behaviors needed for agility. Lure and reward your dog for turning in a circle next to you. Gradually increase speed and size until they circle an object. Add the verbal "go around!" cue once the behavior is consistent.

Contact training accustoms dogs to walking across novel surfaces. Place a board or platform on the ground and lure your dog onto it with a treat. Mark and reward four paw contact. Gradually fade the lure until your dog reliably boards from the verbal "on!" cue.

Jumps should start low, at 25% of the dog's height. Toss a toy over the bar to lure your dog over, marking and rewarding jumping. Gradually increase bar height as your dog masters each increment. Remove the toy lure and add a "jump" voice command once the skill solidifies.

Weaves take patience, as dogs must learn to smoothly enter and change direction between poles. Lean two poles against each other to form an A-frame. Lure your dog through the opening with treats, then gradually separate the poles, marking and rewarding each time they weave properly.

Add cues one at a time for each obstacle. Link behaviors into simple sequences like: touch - go around - jump. Hand targets help guide your dog through the course. Work up to doing full courses of several obstacles in sequence by rewarding generously.

Avoid drilling repetitive practice of the same obstacles - keep things varied and engaging. Read your dog's signals - don't progress too quickly or do too many obstacles to prevent frustration. Staying positive keeps the experience rewarding. Proper shaping builds confidence in novice dogs.

Home training allows you to:

- Habituate your dog to equipment through play and treats

- Establish essential targeting and shaping skills

- Practice handling maneuvers like crosses and sends

- Proof obstacle behaviors through distraction training

- Set a foundation before taking public classes

- Bond further through this interactive activity

Take advantage of your backyard space to acquaint your dog with agility under zero pressure. Home training lets you adapt activities to your unique dog's needs. When they transition to classes, your prep work allows them to focus on novel environments and group work rather than new obstacles. Consistent foundation skills cultivated at home enhance public training success.

Agility provides immense mental and physical enrichment for dogs and handlers. Training together builds communication, trust and athleticism. Home training constructs critical competencies so that when you and your dog step into the competition ring, they're prepared to confidently run the course and excel together.

Training Your Dog to Follow Scent Trails

Teaching your dog to follow and detect specific scents is a mentally stimulating activity that taps into their incredible olfactory abilities. From casual trailing games to competitive nosework trials, scent work provides fun mental enrichment while building confidence and focus in dogs. Where do you start in training your dog for scent trail work?

First, determine your dog's scent drive. Breeds like Beagles, Bloodhounds and German Shepherds are naturally gifted trackers, while less scent-driven dogs may need more coaxing. Test their motivation by hiding treats and rewarding your dog for finding them. Does your dog love sniffing out food rewards? Then their nose is ready to get to work!

Start scent training in a low-distraction indoor area. Lay a short trail of medium-value treats on the floor, spacing them a few inches apart. Encourage your dog to "Find it!" as they approach the trail, then reward them for following the scent and consuming each treat. Gradually increase trail length and decrease treat frequency as your dog masters this introductory tracking game.

Once your dog reliably follows straight indoor trails, create meandering trails by placing turns and weaving paths. Spacing treats a foot or more apart challenges your dog's nose to work harder seeking the next scent clue. Advance to hiding treats under plastic cups along the trail to add a discovery element. Your dog gets mental enrichment from problem solving how to find each treat.

Now introduce outdoor trailing by laying trails in grass using high-value treats like chicken. Start with short, simple paths and reward your dog upon completing the trail. Avoid areas with heavy scents and distractions at first. Over many sessions, build up to longer, more complex trails across various outdoor terrain. Use longer intervals between treats to challenge your dog's persistence while following a scent path.

Once your dog masters basic trailing, advance training by using novel scents. Soak cotton balls, socks or cloth in essential oils like lemon, lavender or eucalyptus. Set scented articles along a trail versus treats. Now your dog must identify and follow a specific smell to find the next clue. Introduce different scents in each session to keep your dog's nose working hard!

You can create fun hide-and-seek games by trailing to a hidden Treat or toy reward. Set longer scent paths winding through rooms or the backyard leading to a plush toy or a stash of treats your dog gets to uncover. This combines mental stimulation with the reward of a treasure hunt!

To proof reliability, conduct training sessions in varied environments - woodlands, parks, beaches, etc. Expose your dog to different terrain, scents and distractions while tracking. Check that they remain focused on the target scent. Doing short practice trails before exercise or play keeps your dog's nose primed.

Use obedience cues like "Find it" or "Go sniff" to focus your dog when first giving them a scent trail. Once they are experienced, allow them to work independently, using their natural instincts to follow the smell to the reward. Let your praise be their guide. Give them freedom, checking in verbally, as they traverse trails off-leash.

Patience is essential, as scent work requires many incremental steps over a long training period. Keep sessions short and upbeat. Troubleshoot issues like lack of focus, frustration, or confusion by revisiting foundations. Regress the difficulty level if your dog seems mentally taxed. Make sure outbound trails lead back to you for safety.

With dedicated, positive training, your dog can progress from following basic food trails to mastering complex scent sequences, distance work, and novel odors. Their mental stamina, odor memory and problem-solving abilities will grow tremendously. From initial curiosity to advanced nosework titles, scent trail training provides limitless mental enrichment possibilities. Most importantly, this activity stimulates natural behaviors in your dog for their greater fulfillment.

Teaching Complex Tricks and Sequences

Stringing behaviors into chains creates dynamic performances. Planning intricate tricks challenges and engages dogs mentally and physically. Breaking sequences into incremental steps lets dogs succeed. Creativity plus patience equals impressive results.

Shaping techniques gradually build new skills by reinforcing approximations toward a goal behavior. Small successes motivate perseverance on incremental progress. Use high-value rewards to maintain motivation on complex tricks. Breaking down each maneuver into its simplest parts enables systematic development.

Leverage previously trained foundational skills when tackling elaborate tricks. Solid obedience, targeting, and behavioral fluency provide the basic alphabet for sequencing. Ensure each individual trick is fluent before chaining in rapid succession. Rushing forward too quickly risks frustration.

Plan multi-step routines in reverse order, teaching the end behavior first. The final trick becomes reinforced as the endpoint, chaining earlier steps on gradually. Backward shaping helps dogs learn the overall sequence more seamlessly. Executing even part of a known pattern earns rewards.

Varying reinforcement schedules is key for long sequences. Randomly provide jackpots just for portions successfully completed to buoy confidence. Maintain engagement with games or easy efforts between complex repetition. End each session on a high note after a success.

Use target sticks, lure sticks, or hand signals to guide positioning through intricacies. Clear cues demonstrate desired movements. Fade prompts gradually as dogs master each segment. Precision handling and timing creates polished performances.

Avoid rushing into overly long or demanding sequences before independent skills are solid. Assembling chains should build on proficiency, not create confusion. Troubleshoot mistakes methodically and rebuild where needed. Patience pays off in the finished product.

Mental engagement comes from learning something new, not just repeating the familiar. Increase difficulty gradually by lengthening sequences, quickening pace, or adding environmental distractors. Vary routines to prevent boredom once mastered.

Learning complicated series requires focus and persistence from both trainer and dog. Celebrate small achievements along the way. Dogs relish the chance to bond and play. Challenging their minds pays cognitive dividends beyond the final applause.

CHAPTER 6

BUILDING A STRONGER BOND

The Importance of Quality Time

Dogs thrive when their human companions dedicate quality time to nurture the relationship. Consistent positive interactions build trust and strengthen bonds. Setting aside periods focused entirely on your dog without distractions conveys the depth of your bond. Quality time benefits both parties.

Dogs are social creatures who crave attention from their loved ones. While existing in the same space provides some companionship, your dog yearns for true quality engagement. Activities done together like walks, play sessions and snuggle time promote oxytocin, the "love hormone". This reinforces attachment.

Quality interactions require your undivided attention. Eliminate smartphones and multitasking when spending dedicated time together. Eye contact, speaking conversationally and touch convey you're 100% present. Following your dog's lead during play or training shows respect.

Create rituals that become cherished routines. A morning walk around the neighborhood or game of tug before dinner brings security through consistency. Rituals give dogs something positive to anticipate. Vary locations and activities to prevent boredom. Dedicate weekends to longer adventures together.

Be fully in the moment instead of just accomplishing obligatory exercise. Treasure simple joys like your dog scampering in fresh snow or sniffing their favorite bush. Share those little miracles that mean so much to our canine friends. Match their playfulness and enthusiasm. Prioritize their preferences sometimes instead of rigid human plans.

Use technology thoughtfully - capture special moments on video or camera to look back on fondly. But avoid viewing your dog's antics through your screen instead of real life. When spending quality time, keep devices put away to eliminate distraction. You'll both gain more enjoyment that way.

Quality time requires sincerely observing your dog's personality - what makes them tick? What do they love most? Customize activities around their unique needs. An elderly arthritic dog benefits more from gentle massages and short strolls vs. marathon play sessions. Learn your dog's language too - how they express happiness, stress or curiosity.

To cultivate quality time:

- Establish set periods devoted solely to your dog without multitasking. Keep it frequent - daily is ideal.

- Engage in mutually enjoyed activities that nurture your bond like cheerful training sessions or meandering walks.

- Truly savor your time together - don't view it as another task to check off.

- Reduce outside stressors and distractions to be fully present.

- Take cues from your dog - let them set the pace and direction sometimes.

- Observe subtleties about their preferences and personality. What strengthens your rapport?

- Capture memories without intrusive technology - be in the moment.

- Maintain routines but sprinkle in novelty too.

Remember - quality supersedes quantity. Brief moments of genuine connection convey more devotion than longer distracted periods. Consistency also matters - sporadic quality time won't fulfill your dog's needs. Prioritizing everyday quality time cements the human-canine bond.

Activities to Enhance the Human-Canine Bond

The unique relationship between humans and dogs is one of the most profoundly special bonds in nature. Enriching that connection through shared activities and quality time together greatly benefits the wellbeing of both species. What are some impactful ways to enhance the human-canine bond?

Simply taking time to play with your dog each day is an excellent bonding activity. Set aside at least 20-30 minutes daily exclusively for playtime. Engage in your dog's favorite games - fetch, tug-of-war, chase, agility obstacles - fueling their natural enthusiasm. Aim for activities that are mutually fun to boost the enjoyment you share. Laughter and silliness strengthen bonds.

Another key way to connect is through training exercises. Even 5-minute sessions practicing commands or tricks create positive shared experiences. Maintain upbeat energy and reward your dog generously. Training builds understanding and communication between you and your dog. Accomplishing goals as a team forges closeness.

Explore new places and have adventures together. Visit novel parks, trails, beaches and other dog-friendly locales. Bring toys and treats to turn excursions into fun bonding outings. Exposure to new sights and smells with you expands your dog's world. Shared discovery cultivates lasting memories.

Pamper your dog with gentle massages, brushing and hand feeding. As you provide these caring touches, look into your dog's eyes and speak softly. Intimate nurturing moments promote trust and contentment. Adding relaxing aromatherapy heightens the Zen!

Engage your dog's mind with interactive toys and food puzzles. Work together manipulating and figuring out brain games. Collaborating to solve mental challenges is rewarding teamwork. It shows your dog you value their intelligence.

Get creative in thinking of activities that align with your dog's interests. Does your dog love to run? Take them jogging or biking. Is your pup toy-motivated? Play stimulating hiding games with their favorite stuffed animal. Tailor activities to your dog's personality and preferences. Shared hobbies build understanding.

Let your dog choose the bonding activity sometimes. Observe what motivates them - is it chasing squirrels? Digging holes? Chewing a certain toy? Then make that your shared playtime. Allowing your dog autonomy to direct your interactions builds confidence in the relationship.

Photograph and video precious moments together. Compile pictures celebrating your adventures, snuggles and smiles. Displaying treasured memories reinforces how much the bond means. Create a photo album or short video set to happy music.

Manage health together through exercising, maintaining a healthy diet and weight, and visiting the veterinarian. Caring for your dog's physical needs improves wellbeing for both of you. Sharing healthy habits strengthens your teamwork.

Train your dog for therapy certification or a Canine Good Citizen title. Going through a structured program together toward a goal can deepen mutual respect and pride in the relationship. Accomplishing something that helps others is fulfilling.

Schedule overnight trips and vacations with your dog when possible. Getting away just the two of you provides valuable focused time and novel shared experiences. Make new memories exploring together.

There are endless options for enriching the human-canine bond each day. Simple consistency in the relationship is key - the small moments of play, training, care and adventure all cultivate closeness. Deep bonds thrive through expressing affection, having fun, supporting growth and enjoying each other's company. Cherish and celebrate your priceless friendship.

Communicating Effectively with Your Dog

Dogs and humans speak different languages. Effective communication bridges this gap by understanding canine perspectives. Observant handling paired with clear consistent signaling optimizes mutual understanding.

Dogs primarily take in information through body language, scent, and tone of voice. Subtle cues like facial expressions, posture, and pheromones provide insight. Direct prolonged eye contact can seem threatening instead of engaging. High pitched vocal tones express friendliness and praise.s rely heavily on verbal instruction and visual cues which dogs don't intrinsically understand. Using canine-centric signals like toys, treats, and physical prompts improves clarity. Body language conveys as much as words, if not more. Patience allows dogs time to interpret meanings.

Observing natural canine behavior provides guidance. Dogs sniff to gather context and approach cautiously when unsure. Crouching with tail tucked signals fear. Play bows invite social interaction. Yawning may indicate stress. Relaxed panting helps moderate excitement.

Mimicking canine postures and proximity can smooth interactions. Crouching side by side feels less imposing than hovering overhead. Gentle touch and soothing tones ease tension. Letting dogs initiate contact respects their preferences. Moving at their pace builds trust.

Timing is critical when conveying approval or corrections. Instantly rewarding desired behaviors links the action with the payoff. Delayed responses confuse. Use marker words like "yes!" to pinpoint precise moments for reinforcement.

Eliminate mixed signals that dilute clarity. Cuing behaviors while unintentionally rewarding the opposite undermines communication. Staying neutral when unwanted actions happen prevents reinforcement. Behaviors strengthen through consistency.

Simplify verbal cues for comprehension. One or two word commands like "sit" or "stay" are more distinguishable than complex sentences. Use an upbeat friendly tone to hold attention. Saying their name before giving instructions focuses dogs.

Visual signals aid verbal cues. Hand gestures, target sticks, lure sticks and body positioning provide clarity.Signals should be unique to avoid confusion between behaviors. Give cues patiently until each is mastered.

Thoughtful body handling adds nuance, guiding dogs gently into proper positions. Harnesses allow better control than leash pressure on the neck. Moving slowly and steadily calms excited energy. Sudden movements can startle and break focus.

Mutual understanding requires effort from both parties. While dogs adapt more to human norms, we must also learn canine subtleties. Blending verbal, visual and tactile communication based on their perspective bridges the translation gap. With patience and empathy, effective partnerships flourish.

The Art of Canine Massage

Massage offers countless benefits for dogs, from reduced anxiety to pain relief. Learning massage techniques to practice on your dog enhances wellness and strengthens your bond. Canine massage utilizes gentle manipulations and pressure to relax muscles, stimulate circulation and soothe joints. An introductory guide to the art of dog massage provides the basics.

Always start slowly and gauge your dog's comfort level. Short, focused sessions allow them to acclimate to handling. Positive associations are key—massage before dinner or pair with treats. Avoid areas of pain or sensitivity. Set a calm environment with minimal distractions.

Warm your hands first and use a flat palm to make broad strokes down the length of your dog's back and limbs. Apply gentle pressure with the whole hand in long, flowing motions. Repeat these relaxing effleurage strokes as your dog settles into the massage.

Kneading motions help release muscular tension. Grasp gently and compress, then release repeatedly using your fingers and thumbs. Target areas like the shoulders, haunches, and thighs where dogs tend to hold tension. Avoid bony prominences.

For compression, apply gentle direct pressure downwards. Use thumbs or knuckles to compress large thigh or shoulder muscles. Increase and decrease pressure in increments to smooth fascia.

Passive stretches provide relief too. Gently extend front and back legs forward, back, and to each side. Avoid over-extending. Stretches improve mobility if done carefully.

Ideally, massage moves proximal to distal. Start at the shoulders/hips, work down the legs, then finish at the toes/tail. Listen for sighs of relief as tight spots release. Notice areas of sensitivity to address gently. Friction knots and circular thumb motions help break down adhesions.

Key pressure points to include:

- Base of tail – releases lower back

- Upper inner thigh – calms nervous system

- Between toes – stimulates organs

- Center chest – aids breathing

- Above eyes – relieves headache

Avoid direct pressure over joints or the spine. Never force range of motion—let your dog move at their comfort. Stop if they seem distressed or nip. Keep sessions brief at first.

Benefits of massage:

- Reduces pain and stiffness, increases mobility

- Lowers anxiety and heart rate

- Releases endorphins and oxytocin for bonding

- Improves circulation and lymphatic flow

- Develops trust and connection through touch

Regular massage enhances your dog's health. Compassionate touch heals both physically and emotionally. As your technique improves, your dog relaxes into your capable hands, receiving relief. Through massage, you "pet with a purpose", nurturing your loyal companion.

The Role of Play in Relationship Building

Play occupies a vital role in building strong relationships between humans and dogs. Playtime provides a positive outlet for dogs to engage their natural instincts and fulfillment. In turn, humans develop trust and an intuitive bond with their dogs through play. Shared play builds confidence, communication and compatibility.

On the canine side, play satisfies innate needs for enrichment. Dogs descend from predators like wolves who learn crucial life skills through play. Puppies play-fight to develop coordination and social rules. Seeking games fulfills inborn drives to hunt and forage. In play, dogs can safely satisfy these impulses in acceptable ways with human partners.

Play also teaches dogs appropriate interaction with humans. A puppy learns how hard to bite or when to stop chasing through play with people. Dogs establish positive habits like retrieving toys instead of nipping hands. Gaining these skills through play makes training easier.

On the human side of the bond, play enables observation of a dog's natural personality. People get insight into attributes like their dog's favorite games, play style, drive, and energy level by playing together. This understanding informs better care and training tailored to the individual dog.

Shared play also forges an intuitive connection. Partners that play together regularly become in sync. Humans learn their dog's body language for when play gets too rough or their pet needs a break. Dogs understand human cues for when to chase a ball or come get a belly rub. Frequent play creates seamless teamwork.

Play builds trust between species. Dogs display vulnerability by exposing their bellies for rubs or allowing handling during play. Safe, positive handling through play accustoms dogs to human touch. A history of mutually enjoyed games establishes confidence in interactions.

Shared laughter and excitement during play strengthens emotional bonds. Playing together just feels good! Releasing feel-good endorphins while having fun is powerful for human and canine. A sense of friendship and loyalty develops when play is consistent.

Setting aside dedicated playtime each day is key. Dogs thrive on routine and expect this cherished ritual. whether it is fetch sessions, tug-of-war, or rhyming lessons, keep favorite games consistent. Spontaneous bursts of play further reinforce the bond.

It's important play remains mutually enjoyable for dogs and humans. Watch for signs your dog is done playing like hiding toys or avoiding interactions. End on a positive note before fatigue or disinterest sets in. Keep things fun, not frantic.

Use play to positively reinforce desired behaviors in your dog like sits, stays and focus. Have your dog "work for play" by requiring a cue or trick before throwing a ball. Linking play with training solidifies skills.

Take playtime on the road! Play games in new outdoor locations to provide mental stimulation while deepening your connection through play. Outdoor exploration fuels dogs' sense of adventure.

Make training feel like play too. Incorporate praise, encouragement, variety and rewards into practice to make it more playful. Your upbeat energy will motivate your dog.

Track your dog's play preferences. Note games, toys and activities they seem to engage with most enthusiastically, then maximize those for bonding playtimes. Customize play to your dog for best results.

Play games that channel your dog's natural drives like hiding kibble around the house to fuel foraging instincts. Or play hide and seek calling them to come find you using their hearing.

Avoid forcing play upon dogs showing disinterest and monitor for signs of overstimulation or anxiety. Forcing interactions can harm trust. Respect your dog's preferences and mood.

Embrace silly, goofy play. Make playtimes full of laughter, bonding and joy. Let your inner child out and observe your dog's natural enthusiasm. Your shared inner happiness will be palpable.

Shared play enables dogs and people to relate on an instinctual level. Play fuels cooperation, deep contentment and mutual understanding. Make play a cherished daily ritual and integral part of your relationship. When dogs and humans unite in play, an unbreakable bond forms.

CHAPTER 7

MENTAL STIMULATION FOR SPECIAL CASES

Mental Exercise for Puppies

Raising mentally engaged puppies establishes lifelong learning. Just like children, puppies thrive through educational play and discovery. Tailoring brain games to developmental stages stimulates growing minds.

The first weeks focus on sensory stimulation - interesting sights, sounds and textures. Novel experiences in small doses expose puppies to the world during imprinting periods. Gentle handling and socialization prevents future anxiety.

As coordination develops, games build physical skills. Fetching toys, tunnels, wobble boards and low obstacles strengthen motor skills with mental challenges. Structured play with siblings teaches critical social bite inhibition.

Three to sixteen weeks is the key socialization window. Positive exposure to people, pets, environments, handling and activities sets lifelong confidence. Continued interactions reinforce early imprinting. Maintaining optimism prevents fear stages.

Basic training can begin as early as seven weeks using positive reinforcement. Short, fun sessions with food lures and rewards tap into fast learning abilities. Foundation cues like look, name recognition, come, sit and stay establish communication.

Young puppies have limited attention spans - 5 minutes or less. Keep lessons simple, rewarding generously to end on successes. Schedule brain breaks to prevent cognitive fatigue. Make everything a game at this age.

Challenge levels should increase appropriately as cognitive skills develop. Vary games and environments to prevent boredom. Introduce new toys and tasks like finding hidden treats to problem-solve. Expand training duration as focus strengthens.

Adolescence from 6-18 months presents new training opportunities as puppies transition to adult abilities. Higher complexity tricks, public access skills and off-leash reliability can be reinforced during this prime learning time.

Preventing destructive chewing is key by providing outlets for mental and physical energy. Food puzzles and interactive toys limit misbehavior. Reinforce calmness and impulse control. Settle and relaxation skills will pay off.

Mental exercise should remain part of the routine even after puppyhood. An engaged mind continues maturing. Learning and quality time strengthen the human-canine bond for a lifetime. Custom games and training keep life exciting.

Early positive experiences and progressive mental enrichment prime puppies for success as adults. Smart socialization and training stimulates neuroplasticity. A strong foundation supports growth into confident, capable companions.

Adapting Games for Senior Dogs

Mental exercise remains important even as dogs age. While senior dogs have different physical and mental needs, games can be adapted to keep them sharp. Adjusting physical and cognitive activities prevents boredom and maintains quality of life. Games for senior dogs tap into wisdom learned over their lifetime.

Focus on low-impact games to avoid joint pain. Hide-and-seek by concealing treats allows light movement without strain. Place tasty snacks under overturned buckets or boxes and have your dog seek the hidden treasure. This utilizes their keen nose.

Snuffle mats or blankets with treats tucked in folds encourage natural foraging instincts with limited mobility required. Adding essential oils creates brain-stimulating scents to identify. Placing kibble inside egg cartons or cardboard tubes challenges cognitive skills too.

Slow controlled games of fetch down hallways or along fences keep senior dogs engaged in play while minimizing excessive activity. Use soft toys instead of balls to prevent mouth pain. Always monitor for signs of fatigue.

Food puzzles like Kongs frozen with wet food or snuffle balls with dry kibble provide cognitive challenges without significant movement. Watching your dog manipulate toys to earn the rewards inside stimulates their mind.

Repetitive training games promote mental sharpness as senior dogs practice previously learned commands. Even basic cues like sit, down, stay and come exercised consistently maintain their skills. Keep training sessions short but frequent.

Adapt walks to your senior dog's needs by going at their slower pace, taking rest breaks and avoiding high-impact surfaces. Vary routes to add mental stimulation. Outdoor exploration remains enriching when tailored appropriately.

Provide comfort like orthopedic beds and ramps to minimize stiffness. Massage increases circulation and relieves arthritic pain. Keeping nails trim prevents slipping during limited activity. Maintain a consistent routine for confidence.

Maintain social and sensory enrichment too. Sniff walks in new environments keep life interesting. Positive interactions with other calm dogs and friendly people prevent isolation. Hearing familiar voices, smells and sounds aids cognition.

Monitor for signs of cognitive decline or anxiety disorders that may require medication. Check vision and hearing regularly as sensory loss impacts mental engagement. Consult your veterinarian about brain health concerns.

The key is discovering age-appropriate games that nurture your senior dog's abilities without overwhelming or frustrating them. Adapt activities to their current interests, mobility and stamina. They'll continue reaping mental benefits from exercises modified for their needs.

With creativity and compassion, you can enrich your older dog's golden years. Simple, low-impact games that spark mental vitality countdownanion. Relishing their slower rhythm lets you appreciate the winding down of a life well lived.

Exercises for Small or Less Active Breeds

While all dogs benefit from mental stimulation, brain games need adjusting for small breeds and less active dispositions. Certain breeds like Chihuahuas, toy poodles, pugs and Cavalier King Charles spaniels tend towards shorter attention spans, lower energy and more delicate physical build. Adapting activities for their size, abilities and temperament ensures effective mental exercise.

Firstly, keep training sessions brief but frequent. Miniature breeds typically have a harder time focusing for prolonged periods. Keep most interactive games or training sessions to 5-10 minutes maximum. End on a positive note before your dog loses interest. With short bursts throughout the day, small dogs still benefit from brain games without overtaxing them.

Lower impact exercises are also key for fragile or tiny dogs prone to injury and exhaustion. Instead of vigorous fetching, have them hunt for hidden treats around a room. Use soft foam ramps and low hurdles for mini agility courses. Introduce food puzzles that require cognitive skill more than physicality. Impact-free activities avoid stress on delicate joints and backs.

Incorporate more scenting and nosework. Let pint-sized pups use their powerful nose to sniff out kibble in snuffle mats, boxes and toys. Track training capitalizes on keen scenting ability without tiring small bodies. Hide-and-seek games likewise employ your small dog's keen sense of smell and hearing versus speed or agility.

Motivate less active breeds with enticing rewards that require little physical effort. Food puzzles and treat-release toys easily capture attention spans of less energetic dogs. The mental challenge of earning tasty food morsels engages their minds with minimal exertion.

Moderately difficult toys are more suitable for diminutive dogs than overly simple or highly challenging ones. While the latter could frustrate them, overly easy toys fail to provide cognitive enrichment. Find interactive puzzles calibrated for your individual dog's problem-solving abilities.

For less agile toy breeds, use lighter toy options appropriate for their size. Avoid heavy ropes and solid rubber fetch balls that strain delicate jaws. Light plush toys and soft flying discs suit them better. Adapt tug-of-war with thin fleece ribbons they can manage versus rope. Customizing to your dog's physical abilities prevents injury.

Take advantage of your small dog's portability to provide mental stimulation in more scenarios. Bring compact training tools like clickers, mini treats and folding platforms on outings for impromptu training sessions. Vary locations more frequently to ignite your portable pup's senses. Exposure to changing environments intrigues them.

Tailor mini agility courses at home using props suited to your dog's abilities. Use low balance beam walkways, short step ladders for climbing, tunnels made from rain gutters, narrow limbo poles low to the ground. Alternate activities regularly to maintain curiosity.

Incorporate lightweight leash walks versus strenuous hikes. Exploring new neighborhoods provides outdoor enrichment for tiny pups without depleting their energy. Let them sniff and engage with safe surroundings while remaining leashed and protected. Bring motivating toys to deploy when attention wavers.

Take education cues from your diminutive dog. Notice when they lose interest and adapt activities to recapture motivation. Observe their unique style of play and interaction to determine ideal games. Small dogs will reveal their preferences.

Factor in any physical limitations present in senior and special needs dogs when tailoring mental exercise. Adjust intensity, duration and difficulty to avoid strain. Basic obedience training provides cognitive stimulation without demanding mobility for arthritic dogs. Avoid frustration by simplifying overly complex tasks based on your dog's abilities.

The small or mellow dispositions of certain breeds requires fine-tuning typical brain games. But simple adjustments like shortening sessions, reducing exertion and customizing to your dog's needs can ensure their mental and physical needs are met. Bring out their inner big dog with brain challenges!

Activities for Large or High-Energy Breeds

Big and driven dogs require adequate physical and mental exercise to prevent destructive behaviors. Tailoring stimulation to match their capabilities channels excess energy constructively.

Brisk walks and jogging provide necessary cardiovascular activity for large breeds. Swimming, hiking and structured running allow dogs to stretch their legs and explore new environments. Group dog sports like agility, dock diving, or lure coursing tap into athleticism.

Mental enrichment is equally important to tire minds. Nosework develops natural scenting ability through finding hidden treats. Vary locations to introduce new mental challenges. Up the ante for proficient sniffers with obstacle navigations.

Herding breeds excel at rally courses with cones, jumps and directional signage. Shepherding tennis balls or toys stimulates herding instincts. Learning new tricks and commands provides mental workouts.

Puzzle toys and food dispensers allow independent play. Adjust opening size and difficulty to extend engagement. Durable chew toys provide healthy outlets for power chewers. Supervise to prevent ingestion of household items.

Training impulse control prevents hyperactivity and jumpiness. "Settle" and "place" commands reinforce calmness and waiting politely. Gradually increase duration and distraction levels to strengthen focus.

Socialization is lifelong for gregarious breeds prone to exuberance when meeting new people. Maintaining manners prevents scary behavior. Crowds offer exposure to desensitize reactions.

Adolescent dogs benefit from advanced training to build skills and challenge energy levels. Developing off-leash reliability expands freedom for exploration. Games maintain engagement as dogs mature.

Mental fatigue is just as real as physical tiredness. Balance strenuous activity with relaxation to avoid stress or injury. Know when high-energy dogs need down time to recharge. Not all days need to be intensely active.

As dogs age, adapt activities for changing abilities and energy levels. Lower impact exercise, food puzzles, locating favorite toys and short training sessions continue providing purpose.

All dogs need constructive stimulation tailored to their capabilities and preferences. Paying attention to each individual prevents boredom and bad behaviors. A fulfilled dog is a well-behaved companion ready to relax and engage as needed.

Mental Stimulation for Dogs with Physical Limitations

When dogs have disabilities or restrictions on their mobility, mental enrichment becomes even more crucial. Physically limited dogs require creative solutions to keep their active minds engaged. Adapting both physical and mental exercise prevents boredom while working within the dog's abilities.

For amputee or paralyzed dogs, take advantage of mobility they do have for light activity. Support wheelchairs, harnesses, or slings allow safe standing, walking or range of motion exercises. Massage improves circulation and joint health. Swimming and hydrotherapy build strength where possible.

Dogs with mobility limitations still enjoy car rides for environmental stimulation. Prioritize scent games - hide treats to sniff out or use puzzle feeders with aromatic foods. Learning nose work skills taps into capable senses when the body has defects.

Provide interactive toys that encourage physical manipulation without taxing mobility. Food dispensing toys, snuffle mats and light tug ropes provide cognitive challenges they can work at. Rotate novel toys to prevent habituation.

Take mental stimulation outdoors on good days. Explore new outdoor spaces at an easy pace adapted to your dog's ability. Vary routes and let them sniff and observe. Outdoor adventures provide enriching sensory stimulation.

Try easy agility obstacles adjusted low to the ground. Weave poles or ramps allow success as your dog tackles exercise within their capability. Mark and reward all efforts, no matter how small.

Continue training cues your disabled dog already knows using hand signals when possible. Dogs take pride in showing off their skills. Practice tricks while calmly laying down. Any positive reinforcement activity strengthens cognition.

Patience and compassion are required - don't over-tire a limited dog. Keep games low-key and sessions brief to avoid frustration. Monitor comfort closely and allow rest as needed. Don't force activity that causes pain or anxiety.

Adapt walks for blind, deaf or anxious dogs - uselandmarks, scents, textures and routines to build confidence in navigation. Teach cues like targeting hands for guidance. Verbal praise, vibration or touch signals provide feedback they can perceive when senses decline.

Puzzle toys, snuffle mats, gentle training games and sensory outings give disabled dogs purpose. Adjust all physical and mental enrichment to align with your dog's abilities and needs. They'll continue gaining joy and fulfillment through activities carefully tailored for their situation.

Adapting to Your Dog's Changing Needs

Our canine companions go through various life stages from exuberant puppyhood to mellow senior years. Their mental exercise needs inevitable evolve as your dog matures physically and cognitively. Remaining attentive and adapting brain games to your dog's changing abilities and needs is key to providing mental enrichment across their lifespan.

As puppies, short 5-10 minute training sessions suit limited attention spans but aid critical cognitive development. Simple commands like sit, stay and come establish foundations for lifelong learning. Food puzzle toys teach basic problem solving. Interactive games like tug-of-war and fetch build bonds with human caretakers. Keeping things positive, short and not overly challenging prevents frustration for puppy brains.

The adolescent period from 6 months to 2 years old brings new training challenges. Impulse control and manners may lapse without productive mental outlets. Incorporate more complex commands like weaving through legs or crawling. Vary environments and social situations to provide mental novelty. Praise focused attention and calm behavior. Teens needs consistency combined with increasing mental challenges.

In a dog's prime adult years, more advanced training reinforces mental acuity. Agility courses, scent detection work, elaborate tricks and other intense stimulation strengthens cognitive abilities. Vary games frequently to prevent boredom. Daily obedience drills maintain skills. Interactive toy puzzles prevent restlessness. Their exuberance benefits from plenty of fulfilling mental work.

As senior dogs pass age seven, mental exercise slows to avoid fatigue but remains crucial for brain health. Keep training upbeat but simplify commands, breaking them down into smaller steps. Try food puzzles with adjustable difficulty to maintain cognitive challenge without frustration. Mix short training bursts with rest breaks. Mental exercise sustains senior dogs' engagement and purpose.

Pay close attention as your dog reaches new life stages. Tell-tale signs like shorter attention spans, acting "too old" for games they once loved, or struggling with commands that were easy indicate your dog's needs are evolving. Meet these changing requirements with sensitivity.

Resist rapid transitions between life stages by gradually adjusting mental stimulation over time. Increase difficulty and duration of activities bit by bit rather than all at once. Watch for signs of anxiety, frustration or exhaustion. Let your dog set the pace.

Take cues from your dog when adapting brain games to their abilities. Notice their energy level, enthusiasm and comprehension to determine ideal difficulty and duration for their stage of life. Let your dog tell you what suits them rather than sticking rigidly to routines that are no longer working.

Get creative in adapting activities to physical or mental limitations as your dog ages. For example, arthritis can make sitting difficult, so reward standing focus during short training bursts. If eyesight declines, attach bells to toys for audible fetch. Adapt before opting out of cherished games.

As puppies mature, resist overwhelming them with too many high-level training techniques too quickly. Stick to positive reinforcement and foundational skills. Once basics are mastered, advance to more complex mental challenges. Proper pacing prevents pushing growing brains too hard.

For senior dogs, fight mental decline by keeping them learning. Teach modified old tricks to keep cognitively engaged. Increase treat incentives and simplify puzzle toys to account for any cognitive decline. New sights, sounds and experiences stimulate older brains.

Mental exercise should align with physical activity levels at each life stage. Gradually increase physical demands along with mental challenges as puppyhood transitions to adulthood. Then reduce excessive physical exertion for seniors while maintaining cognitive engagement.

Consistent daily mental stimulation prevents backsliding in already mastered skills. Even short 5-minute refresher sessions reinforce learned behaviors and commands extending their retention. Keeping their brains exercised maintains abilities.

Adapting mental enrichment to your dog's changing needs takes attentiveness and flexibility. But the payoff is a dog who stays fulfilled, engaged and bonded with you across their lifespan. Your adaptable training routine keeps delivering mental health benefits year after year.

Combating Cognitive Decline in Senior Dogs

As dogs enter their senior years, gradual cognitive decline can occur, resulting in disorientation, forgetting commands, and other neurobehavioral changes. Just as physical activity slows with age, the brain's processing power decreases. However, there are many ways to combat mental deterioration in aging dogs and preserve their cognitive abilities through tailored brain stimulation.

Incorporating daily mental enrichment becomes particularly key for senior dogs. Learning new tricks, problem solving with food puzzles, scent detection games, agility activities scaled to their abilities - all strengthen neural connections in aging brains. As long as seniors are healthy, keeping them mentally active slows cognitive decline.

Simplifying formerly mastered behaviors helps re-solidify memory. Break previously learned tricks into small steps again as if re-training. Increase repetition of basic commands to sharpen recall. Frequent mini training sessions strengthen retention versus long, complex demands. Recapping foundations maintains mental sharpness.

Introduce new sights, sounds and experiences carefully to stimulate aging minds. Change up walking routes to explore new neighborhoods. Try safe new surfaces like walking over pebbles or wood chips. Exposure to gentle novelty in small doses keeps senior brains plastic and curious.

Adapting a senior dog's environment also supports cognition. Use baby gates to block off unused rooms and reduce unnecessary stimulation. Ensure good lighting, eliminate glare, and reduce background noise that could disorient an aging dog. Keeping their spaces simplified and calm aids cognition.

Leverage food motivation to engage a senior dog's mind. Most older dogs remain driven by food treats. Using high-value rewards when re-training known behaviors or introducing new games incentivizes mental work. Praise effusively for effort to boost confidence along with treats.

Even just basic obedience drills like sit-stay cues during activities help aging dogs practice focus and response. Integrate 5-minute mini training sessions into their daily routine - before meals, during walks, while brushing. Keeping an active training dialogue strengthens mental stamina.

Food puzzle toys designed to release treats or kibble as dogs manipulate them provide invaluable cognitive stimulation. Choose toys keyed to your senior dog's current abilities and interest level. Adjustable difficulty prevents frustration while still being challenging.

Scenting and nosework games leverage a senior dog's keen sense of smell which remains active in old age. Have them search for treats hidden around furniture or track scented ropes. This taps into natural foraging skills in a low-impact way.

Maintain social and sensory enrichment through gentle interaction with people of all ages, safe play with other compatible dogs, and varying textures to feel. Socialization and sensory input aid cognition as long as properly introduced based on ability.

Enrich their environment with calming stimuli like soft music, audiobooks, and pheromone plug-ins. Soothing, positive ambient stimuli may aid senior cognition and prevent anxiety. Massage calms and comforts stiff bodies while enhancing circulation.

Stay alert to cognitive changes and adjust activities to match current abilities. Increase rewards and simplify requests if your senior dog seems confused by previously mastered behaviors. Shift focus away from failing skills onto enhancing remaining strengths.

Develop a routine of mental enrichment but remain flexible. Sticking to a schedule gives security but allow good days and bad days based on energy levels. End sessions before fatigue sets in. Monitor for signs of stress like panting or avoiding games.

With patience and optimism, you can support quality of life for your senior dog despite inevitable decline. Dedicate time to daily cognitive stimulation tailored to their needs. Maintain their established skills while intelligently introducing novelty. Reward small successes. Adapt games before abandoning play. Together you and your senior dog can keep enjoying daily enrichment to the fullest.

BONUS 1

AUDIOBOOK

Scan the QR code and download the audiobook

Laurel Marsh

BONUS 2

MUSIC FOR DOG

Scan the QR code and let your dog listen to music

Laurel Marsh

EXCLUSIVE BONUS

3 EBOOK

Scan the QR code or click the link and access the bonuses

http://subscribepage.io/kUpl2M

Laurel Marsh

AUTHOR BIO

LAUREL MARSH

Laurel Marsh, an author whose life is woven into the delightful realm of dog sitting. Laurel's journey began as an ardent animal lover, drawn to the joy and companionship that dogs bring into our lives.

By day, Laurel is a seasoned dog sitter, providing a haven for furry friends in need of care and attention. Her expertise in understanding canine behavior, combined with a genuine love for animals, has established her as a trusted figure in the pet care community.

Laurel's days are filled with tail wags and wet noses, a testament to her commitment to the well-being of every four-legged companion she welcomes. Beyond the daily routines of walks and playtime, she has developed a keen understanding of each dog's unique personality, tailoring her approach to create a comfortable and loving environment.

Laurel Marsh

Printed in Great Britain
by Amazon